An Atlas of

MICROBIOLOGY OF THE SKIN

THE ENCYCLOPEDIA OF VISUAL MEDICINE SERIES

An Atlas of MICROBIOLOGY OF THE SKIN

Edited by

Jack L. Lesher, Jr, MD

Professor and Chief of Dermatology, Section of Dermatology,
Medical College of Georgia, Augusta, GA

with contributions from

Raza Aly, PhD

Professor of Dermatology and Microbiology, University of California, San Francisco, CA

Dennis E. Babel, PhD

Assistant Professor of Microbiology, College of Human Medicine &
College of Natural Sciences, Michigan State University, MI

Philip R. Cohen, MD

Associate Professor, Department of Dermatology, University of Texas –
Houston Medical School, and Department of Medical Specialties (Section of Dermatology)
University of Texas, M.D. Anderson Cancer Center, Houston, TX

Dirk M. Elston, MD

Chief, Dermatology, Wilford Hall Medical Center, San Antonio, TX

Foreword by

Kenneth J. Tomecki, MD

Department of Dermatology, Cleveland Clinic Foundation
Cleveland, OH

The Parthenon Publishing Group

International Publishers in Medicine, Science & Technology

NEW YORK LONDON

Library of Congress Cataloging-in-Publication Data
An atlas of microbiology of the skin / edited by Jack L. Lesher : with contributions from Raza Aly ... [et al.] : foreword by Kenneth Tomecki.
p cm. -- (The Encyclopedia of visual medicine series)
Includes bibliographical references and index.
ISBN 1-85070-904-1
1. Skin--Infections. 2. Skin--Microbiology. 3. Skin--Infections--Atlases. 4. Skin--Microbiology--Atlases. I. Lesher, Jack L. II. Aly, Raza. III. Title: Microbiology of the skin. IV. Series.
[DNLM: 1. Skin--microbiology atlases. 2. Skin Diseases. Infectious--diagnosis atlases. WR 17 A88195 1999]
RL201.A85 1999
6'16.5--dc21
DNLM/DLC
for Library of Congress 98-50050
CIP

British Library Cataloguing in Publication Data
An atlas of microbiology of the skin. – (The encyclopedia of visual medicine series)
1. Skin – Microbiology
I. Lesher, Jack L.
616.5'01
ISBN 1-85070-904-1

Published in the USA by
The Parthenon Publishing Group Inc.
One Blue Hill Plaza
PO Box 1564, Pearl River
New York 10965, USA

Published in the UK and Europe by
The Parthenon Publishing Group Limited
Casterton Hall, Carnforth
Lancs. LA6 2LA, UK

Printed and bound in Spain
by T.G. Hostench, S.A.

Contents

Foreword 7

Preface 8

Section I Bacteria 9
by Raza Aly

Chapter 1 Laboratory methods 10
Chapter 2 Staphylococci 15
Chapter 3 Streptococci 20
Chapter 4 Corynebacteria and Propionibacteria 25
Chapter 5 Gram-negative bacteria 29
Chapter 6 Other cutaneous bacteria 34
Chapter 7 Mycobacteria 39

Section II Ectoparasites, arthropods and pinworms 45
by Dirk M. Elston

Chapter 8 Arthropods and human disease 46
Chapter 9 Arachnida 48
Chapter 10 Chilopoda and Diplopoda 56
Chapter 11 Insecta 57
Chapter 12 Helminthic pathogens (worms) 65

Section III Fungi 67
by Dennis E. Babel

Chapter 13 Laboratory methods in mycology 68
Chapter 14 Dermatophytes 71

Chapter 15 **Non-dermatophytic hyaline and dematiaceous molds** **86**
Chapter 16 **Yeasts** **98**
Chapter 17 **Dimorphic fungal pathogens** **103**

Section IV **Viruses** **109**
by Philip R. Cohen

Chapter 18 **Herpesviruses: Herpes simplex virus, varicella–zoster virus and cytomegalovirus** **110**
Chapter 19 **Molluscum contagiosum** **120**

Index **125**

Foreword

We shall not cease from exploration
And the end of all our exploring
Will be to arrive where we started
And know the place for the first time

T.S. Eliot
Little Gidding, 4 Quartets

The world of microbes can be daunting and challenging to most clinicians – new and seemingly myriad organisms, confusing terminology, changing epidemiology and patterns of disease, etc. Infectious diseases, with their therapeutic demands and occasional resistance to therapy, can dramatically affect human health, often with significant morbidity and mortality; indeed, they directly account for one-third of all deaths world-wide. To wit, most physicians need to know some microbiology *vis-a-vis* clinical disease. Dermatologists, a unique breed with special needs, now have a solid reference text which adequately addresses their microbial needs and nicely bridges the gap between the laboratory and the clinical arena.

Jack Lesher and his colleagues have produced a handsome atlas – a striking and colorful compendium of the microbial world as it relates to skin disease. The book is scholarly, yet practical and clinically germane; appropriately current and contemporary; comprehensive, without being encyclopedic; and, perhaps most importantly, very user/reader friendly (a key attribute). As presented, this atlas should be a welcomed resource for dermatologists, microbiologists, and residents in training. I can heartily recommend it.

What more can I say? Turn the pages and see for yourself. You won't be disappointed.

Ken J. Tomecki, MD
Cleveland Clinic Foundation
Cleveland, Ohio

Preface

A common and pertinent question asked by authors, readers and publishers when first confronted with a text is whether there is a need or a purpose for such a book to have been written. Our answer to this question is a resounding "Yes!", and we hope that you will agree that this atlas occupies a unique place among the currently available texts dealing with clinical microbiology. Despite the already large number of sources of clinical microbiology, there is not a single source that covers all of this information in the form of an all-in-one atlas as does ***An Atlas of Microbiology of the Skin***.

Although the primary intended audience for this volume is the dermatologist and the primary-care physician in both academic and private clinical practice, we also consider this atlas to be an invaluable teaching and learning resource for residents in training, particularly those who are involved or interested in the fields of dermatology, primary care and infectious diseases. By focusing attention on the major organisms encountered in bacteriology, ectoparasitology, mycology and virology, we have provided a clear, concise and useful depiction of these organisms in a format that is readily available to clinicians in their office laboratories. In addition, this atlas will serve well as a primary resource for the identification of organisms in terms of gross, microscopic and culture appearances.

Each of the four sections in this book provides basic background information on the use of smears, stains and cultures for the organisms as well as the appropriate means for collecting clinical specimens. Specific organisms are described and accompanied by photographs for further clarification. Clinical photographs are included with correlation of the clinical manifestations with the laboratory appearances of the organisms.

Section I reviews gram-positive and gram-negative organisms as well as a number of other cutaneous bacteria, including mycobacteria. In Section II, the clinical problems caused by ectoparasites, arthropods and worms are described and illustrated. Section III outlines fungal morphology *in vivo*, various stains and fungal culture media along with diagrams of sporulation characteristics and the nutritional requirements of various dermatophytes, followed by the clinical, macroscopic and microscopic features of various dermatophytes, non-dermatophytes, hyaline and dematiaceous molds and yeasts, and a chapter on dimorphic fungal pathogens. Section IV includes the most frequently encountered viruses in clinical practice, namely, the herpesvirus family and molluscum contagiosum, and photographs of their clinical and microscopic features.

This unique and useful atlas will be an important addition to your clinical practice and office laboratory. This book is not, however, intended to sit on your bookshelf, but should be out on your laboratory bench, where it will be readily available for frequent use on a daily basis.

Jack L. Lesher, Jr
Augusta

Section I Bacteria

by Raza Aly

Chapter 1	Laboratory methods	10
Chapter 2	Staphylococci	15
Chapter 3	Streptococci	20
Chapter 4	Corynebacteria and Propionibacteria	25
Chapter 5	Gram-negative bacteria	29
Chapter 6	Other cutaneous bacteria	34
Chapter 7	Mycobacteria	39

Chapter 1 Laboratory methods: Smears, staining and cultivation

The morphological detail of bacteria in their natural state is difficult to observe because bacteria are usually transparent when examined under the light microscope. Several stains allow visualization of different bacterial morphological structures. These include Gram's, metachromatic granular, acid-fast and simple stains. Other specialized staining methods to demonstrate the capsule, flagella and spores are discussed in detail in editions of the ***Manual of Clinical Microbiology***, published by the American Society of Microbiology in 1995.

The most commonly used stains are salts. Basic stains consist of a colored cation with a colorless anion, such as methylene blue^{+} and chloride^{-}. Acidic stains have a colorless cation and a colored anion, for example, sodium^{+} and eosinate^{-}. Bacterial cells have negatively charged phosphate groups of nucleic acid which combine with positively charged basic dyes. In negative staining procedures, acidic dyes are used to stain background material a contrasting color.

It is important to use fresh cultures for routine staining procedures. Old cultures lose their affinity for most dyes; thus, in such cultures, gram-positive organisms may appear to be gram-negative.

Smear preparation and Gram's-stain methods

The smear is prepared by spreading a loop containing broth culture upon the appropriate area of a clean glass slide. Alternatively, a small portion of the colony is taken from a slant or a culture plate, emulsified in a drop of water and spread over the desired area (Figure 1.1). The smear should always be dried in the air (room temperature), then heat-fixed gently over a Bunsen flame to promote adherence of bacteria to the slide. Overheating should be avoided as this will cause distortion of the bacteria. The slide is then cooled to room temperature before staining.

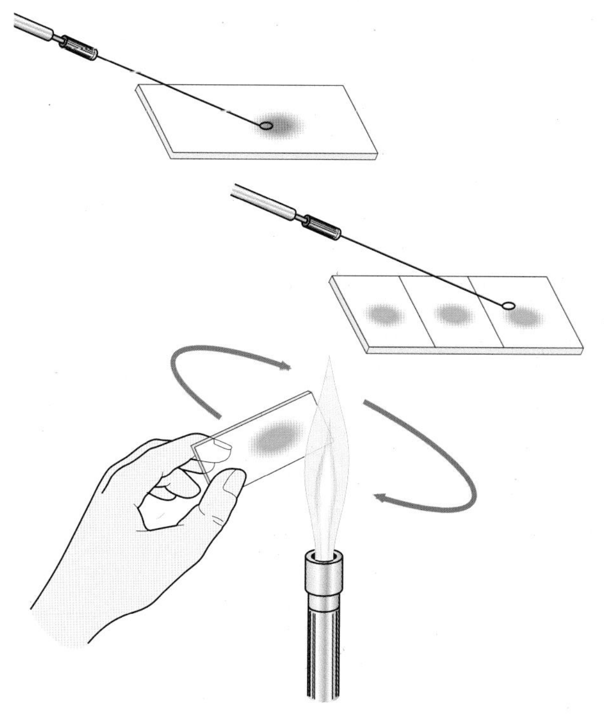

Figure 1.1 Smear preparation

The most commonly used stain in bacteriology is Gram's stain. The procedure is as follows:

(1) Prepare the smear and briefly fix with heat;

(2) Flood the specimen with crystal violet solution for 10 s, then wash with tap water;

(3) Flood the specimen with Gram's iodine solution (mordant) for 10 s, then wash with tap water. All bacteria are stained blue at this time;

(4) Decolorize with ethanol-acetone solution or 95% ethyl alcohol until the solvent flows colorlessly from the slide, then wash with tap water. Gram-positive bacteria retain the crystal violet–iodine complex and remain blue; gram-negative bacteria are completely decolorized by the alcohol;

(5) Counterstain with safranin for 10 s, then wash with tap water. The decolorized gram-negative cells will accept the counterstain;

(6) Blot or air-dry (the latter is preferable).

The critical factor in this procedure is the decolorization step. With overdecolorization, gram-positive organisms may resemble gram-negatives.

When examined under an oil-immersion lens, gram-positive bacteria, which retain the crystal violet–iodine combination, appear purple whereas gram-negative bacteria, which do not retain this combination, appear pink or red due to the counterstain (safranin). The differing results with Gram's staining are due to basic differences in surface structure between these two types of organisms, as confirmed by electron microscopy.

Culture

Most pathogenic bacteria grow on artificial media. Selecting the proper medium for certain fastidious organisms is an important consideration. The culture medium should contain the essential nutrients, and the proper salt and water concentrations. Other factors that must be controlled are the pH and osmotic pressure of the medium. Dehydrated culture media should not be used. Most culture media should be stored in a refrigerator to avoid dehydration, and the media warmed to room temperature before use.

The ability to select a pure colony (a single isolated colony) is the basis of current bacteriological practice. The following technique may serve as a guideline for inoculation of any agar plate.

The swab containing the specimen is rolled completely over one-quarter of the plate surface so that the entire side surface of the swab contacts the medium (Figure 1.2). Similarly, a wire loop carrying the specimen may be streaked across one-quarter of the plate surface. The bacteria may then be spread over the entire plate by touching the inoculated area once or twice with a sterile loop and

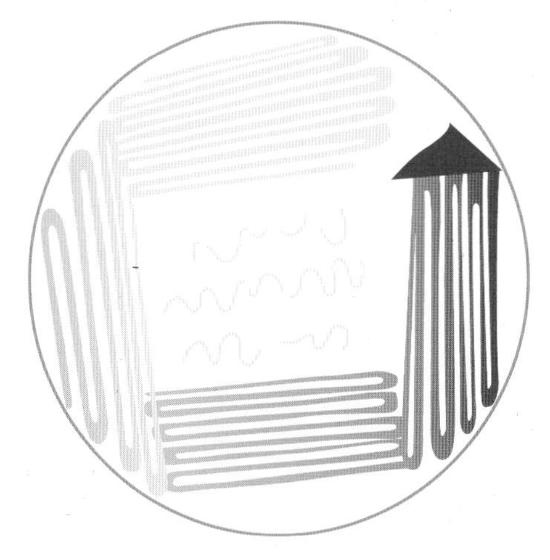

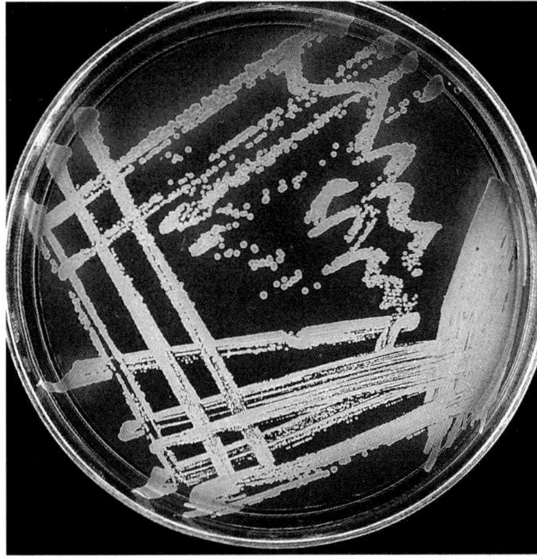

Figure 1.2 Correct technique for streaking a plate (left); streaked plate showing areas of isolated colonies (right)

streaking it across the opposite quarter of the plate. Turning the plate again by 90°, the plate is streaked again using the same loop, and the process repeated until the entire plate surface has been streaked. Single isolated colonies should be seen in the final streaked areas after the appropriate incubation period (see Figure 1.2).

Incubation

Most bacteria of medical significance grow at 35–37°C. Ideally, all cultures should be inoculated in a carbon-dioxide (5–10%) atmosphere. In some laboratories, this may not be possible and cultures are incubated in air. Cultures for meningococci, gonococci and brucellae must be incubated in a carbon-dioxide atmosphere in a candle jar or special incubator.

Anaerobiosis

Anaerobic infections are commonly overlooked because of lack of awareness of these organisms and the more tedious techniques involved in their culture. However, simplification of anaerobic techniques will enable small laboratories to isolate many of these anaerobes and to characterize them with reasonable accuracy. Anaerobic bacteria require reduced oxygen tension and do not grow on solid media even in an atmosphere of 10% carbon dioxide.

GasPak® method

This anaerobic method [Becton Dickinson Microbiology Systems (BBL®), Cockeysville, MD] eliminates the need for the gas cylinder vacuum pump and manometer apparatus required by other techniques. The GasPak (Figure 1.3) consists of a disposable hydrogen-and-carbon-dioxide-generating envelope, and a disposable anaerobic indicator (methylene blue) to create anaerobiosis. The inoculated media are placed in the jar together with one hydrogen-generating envelope, the top corner of which is cut off, and the anaerobic indicator. A pipette is used to add 10 ml of water to the envelope. The cover is then immediately placed in position, and the clamps applied and screwed hand-tight. The hydrogen generated in the jar reacts with the oxygen in the presence of the catalyst to achieve an anaerobic environment. Plastic disposable containers to replace the jar are available (see Figure

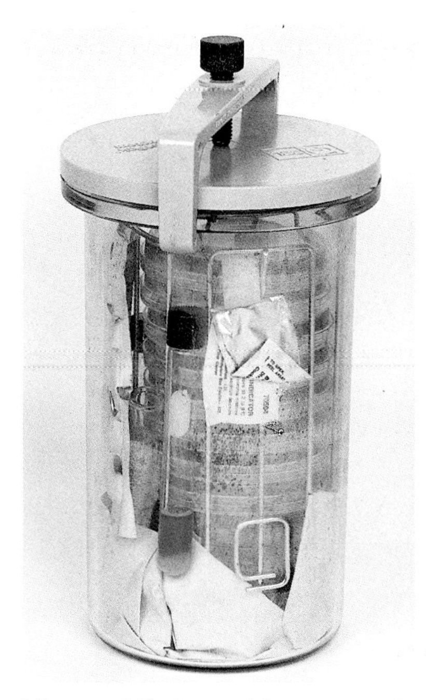

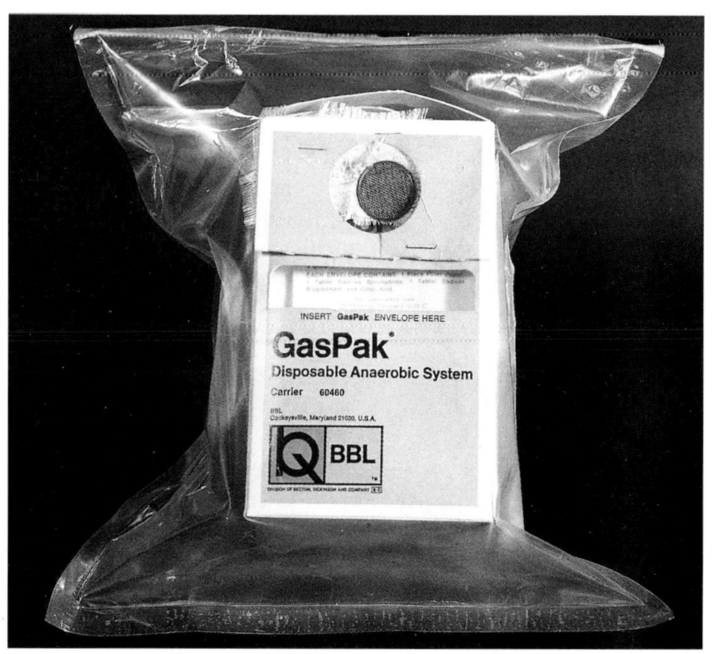

Figure 1.3 Anaerobic system: Anaerobic jar containing a disposable hydrogen-and-carbon-dioxide generating envelope and a disposable anaerobic indicator (left); the plastic disposable container (right)

1.3). After overnight incubation, the methylene blue indicator should appear colorless, thus indicating that anaerobiosis has been achieved. Other anaerobic methods, such as roll tubes and the anaerobic chamber (glove box), are not practical for the average clinical microbiology laboratory.

Media

A critical step in the cultivation of bacteria is the selection of proper media for a specific purpose. Nearly all commercially available media come with instructions and recommendations for use for cultivating particular organisms. Several plate-media preparations are available from BBL or from Difco (Detroit, MI). Terms such as 'enrichment', 'selective' and 'differential' are used to describe certain culture media (Table 1.1).

For general use, blood agar plates (preferably 5% defibrinated sheep's blood) are recommended. This medium is highly nutritive, allows growth of several different kinds of organisms, and serves as a base for hemolytic reactions of aerobes and anaerobes. In many situations, a selective medium combined with a general purpose medium is useful; for example, *Staphylococcus aureus* may overgrow *Streptococcus pyogenes* in blood agar medium when both are present. When crystal violet (1 μg / ml) is added to blood agar, *S. pyogenes* is selected for growth over *S. aureus*. Blood from different animal species is used to demonstrate hemolysis; different organisms give different responses, depending on the blood source.

Chocolate blood agar

This medium contains added blood or hemoglobin and has been heated until it appears chocolate in color. It is particularly recommended for isolation of *Neisseria gonorrhoeae*. The addition of antibiotics such as vancomycin, colistin or nystatin – as in Thayer–Martin medium – renders it selective for *Neisseria* species.

Antibiotic-sensitivity tests

The antibiotic paper-disk diffusion procedure (disk method) is the most widely used antibiotic-sensitivity test; the Bauer–Kirby method has gained widespread acceptance and is recommended. The procedure has been standardized by comparing the size of the inhibition zone with the dilution-test method and with blood levels (normal dosage schedule).

Procedure

Using a wire loop, four or five colonies are transferred to a tube containing 4–5 ml of trypticase soy

Table 1.1 Primary, differential and selective media

Media	*Microorganisms*
Blood agar (prepared by adding blood into brain–heart infusion agar or trypticase soy agar)	Most pathogenic organisms grow on this general-purpose medium
MacConkey agar or eosin–methylene blue (EMB)	Selective and differential media for members of the Enterobacteriaceae family (gram-negative rods)
Thioglycollate broth	Anaerobic bacteria
Mannitol salt agar	*Staphylococcus aureus*
Crytal violet–blood agar	Hemolytic streptococci
Brain–heart infusion agar with 1% glucose added	Anaerobic propionibacteria (corynebacteria)
Thayer–Martin (chocolate agar)	*Neisseria gonorrhoeae*
7H10 and Löwenstein–Jensen (egg-based)	Mycobacteria

broth. The broth is incubated at 36°C for 2–5 h. Turbidity is adjusted by dilution to a barium–sulfate standard (add 0.5 ml of 1.75% w / v barium chloride hydrate to 99.5 ml of 1% w / v sulfuric acid). The bacterial suspension is streaked evenly, using a cotton swab, onto a Mueller–Hinton plate in several directions. After around 5 min, the disks are placed on the inoculated agar surface and gently pressed into the agar with a sterile forceps. The plates are inverted and incubated at 35–37°C. After 16–18 h of incubation, the plates are examined, and the diameters of the complete inhibition zones measured and compared with published correlates.

Bibliography

Beveridge TJ, Davies JA. Cellular response of *Bacillus subtilis* and *Escherichia coli* to the Gram stain. *J Bacteriol* 1983;156:846–58

Isenberg HD, Washington II, Doren GV, Amsterdam D. Specimen collection and handling. In: Balon A, Housler WJ, Hermann KL, *et al*., eds. *Manual of Clinical Microbiology, 5th edn.* Washington, DC: American Society of Microbiology, 1991

Chapter 2 Staphylococci

Staphylococci are currently classified according to specific combinations of phenotypic characteristics and DNA relatedness. At least ten different *Staphylococcus* species are acknowledged to live on human skin: *S. aureus*, *S. epidermidis*, *S. hominis*, *S. haemolyticus*, *S. capitis*, *S. warneri*, *S. saprophyticus*, *S. cohenii*, *S. xylosus* and *S. simulans*. According to the new classification, *S. epidermidis* and *S. hominis* are the most frequently isolated staphylococci from the head, axillae, legs and arms.

Staphylococcus aureus

This organism produces a variety of clinical manifestations, ranging from the simple pustule of impetigo to septicemia and death. The most frequent sites of colonization are the anterior nares and the perineum. Of the general population, 25–45% carry coagulase-positive staphylococci on the nasal mucosa or perineum.

Morphology

Staphylococci are spherical, and range from 0.8–1.0 μm in diameter. They occur in irregular clusters but, in liquid culture media, single cells, pairs and chains are also seen. These organisms are gram-positive (Figure 2.1).

Collecting specimens

Specimens are taken using a number 15 or 11 scalpel blade, or by swabbing the involved areas of the skin. The pus or exudates are spread as thinly as possible on a clean glass slide for Gram's staining. Direct microscopic examination is generally indicated. It is not possible to distinguish other gram-positive cocci from *S. aureus* at this stage. When pus is obtained from an intact pustule, bulla or abscess, demonstration of organisms in biological fluids strongly suggests a pathogenic relationship (Figure 2.2). Culture allows complete identification, but is not a substitute for direct microscopy. The morphological arrangement of individual cells forming irregular clusters in large numbers suggests staphylococci. Culture is required to make the definite identification of *S. aureus*. The other major pathogenic cocci from which it must be differentiated are streptococci. Staphylococci are catalase-positive whereas streptococci are catalase-negative.

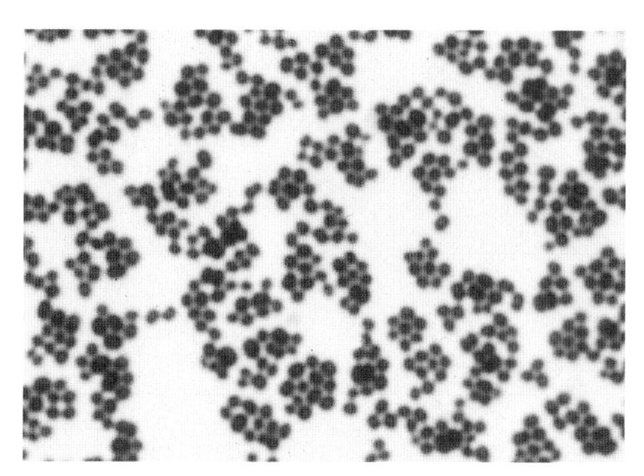

Figure 2.1 *Staphylococcus aureus* seen in clusters (Gram's stain)

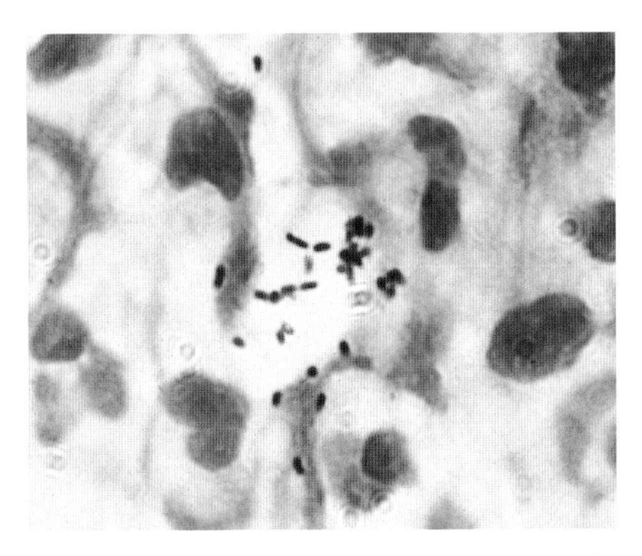

Figure 2.2 *Staphylococcus aureus* from pus. Note the presence of gram-positive cocci (Gram's stain)

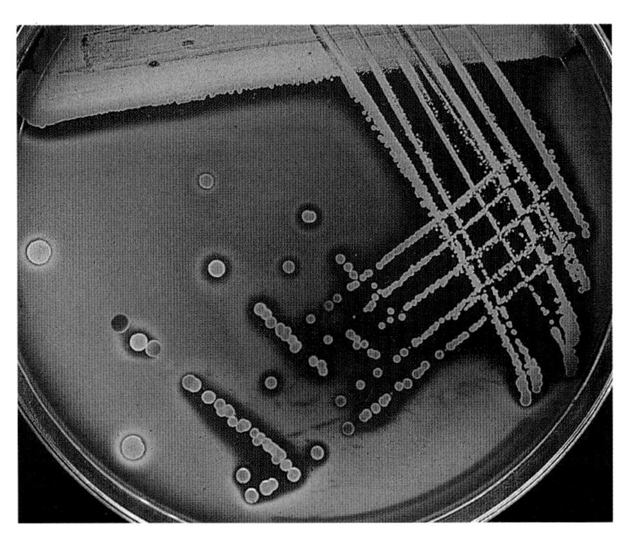

Figure 2.3 *Staphylococcus aureus* on blood agar showing a clear zone of hemolysis around each colony

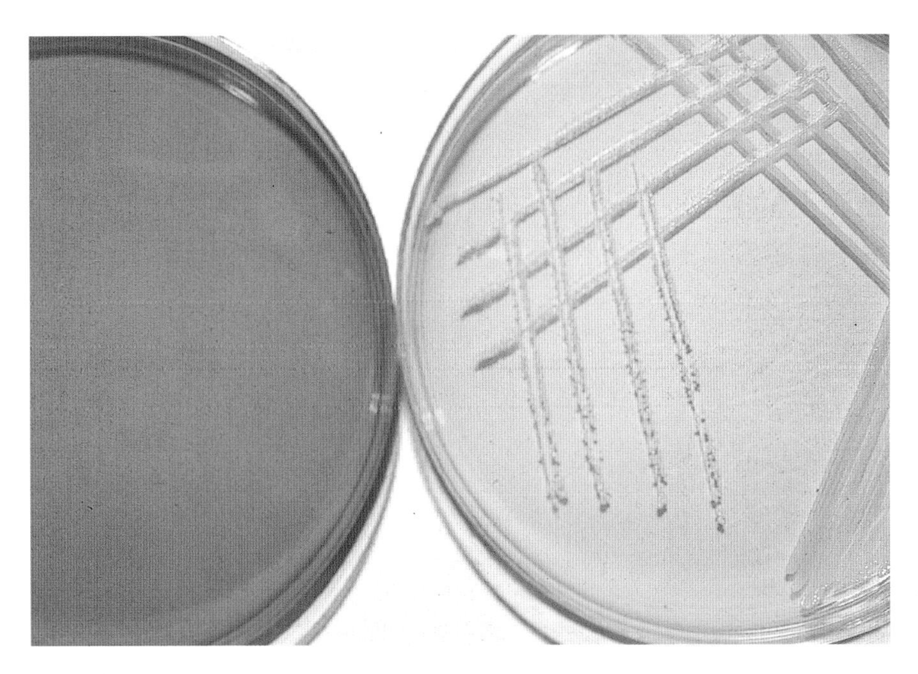

Figure 2.4 Mannitol salt agar plates: Uninoculated (left) and inoculated with *Staphylococcus aureus*, indicated by the appearance of mannitol fermentation (right)

Culture and identification

Staphylococci grow well within 24 h on most general bacteriological media under aerobic and microaerophilic conditions. The optimal temperature is 35° C. Colonies on the agar surface appear round, smooth and glistening. *Staphylococcus aureus* forms creamy to golden-yellow colonies, hence the name of the species; however, colony appearance may vary and is not in itself diagnostic. Pigment production due to lipochrome is enhanced at room temperature. On blood agar, colonies are normally hemolytic and are surrounded by a clear zone of hemolysis (Figure 2.3). Although blood agar plates are sufficient to isolate *S. aureus*, mannitol salt agar as a selective medium is recommended when contamination by other organisms is frequent. The appearance of a yellow color around the colonies is an indication of mannitol fermentation (Figure 2.4), a feature that helps to distinguish *S. aureus* from *S. epidermidis*.

Coagulase tests

The most useful index of pathogenicity is the ability to coagulate plasma; almost all virulent strains are coagulase-positive. Coagulase reacts directly with prothrombin to form a complex called staphylothrombin, which converts fibrinogen to fibrin. The coagulase test is currently considered to be the best available laboratory method for determining the potential pathogenicity of staphylococci.

Slide coagulase test

This easily performed test consists of suspending part of a colony in a drop of water on a glass slide, then adding a drop of human plasma to this suspension. When coagulase is present, the bacteria clumps. Negative results of suspicious isolates may be verified by the tube method (see below). Commercial latex agglutination tests are also available.

Tube coagulase test

Citrated rabbit (or human) plasma, diluted to a ratio of 1:4 in 0.4 ml of water, is mixed with an equal volume of the broth culture of the organism and incubated at 37°C in a water bath. A tube of plasma mixed with sterile broth is included as a control. Complete or partial coagulation within 1–4 h is interpreted as positive.

Catalase test

The catalase test is performed by adding 0.5 ml of 3% hydrogen peroxide to colonies on plain agar slants or plates without blood. Production of gas bubbles indicates a positive test.

Phage typing

Phage typing is invaluable in staphylococcal epidemics and in evaluating recurrent furunculosis. Time-consuming and technically difficult, however, it is performed by few laboratories. Phage typing may thus be obtained by forwarding a pure culture of *S. aureus* in any transport medium to the reference laboratory.

Most strains responsible for staphylococcal toxic epidermal necrolysis belong to phage group II; most staphylococcal hospital epidemics in the 1950s and 1960s involved the type 52, 52A, 80 / 81 complex (group I). The production of enterotoxin is confined primarily to phage groups III and IV.

Bacterial metabolites

Staphylococci release numerous extracellular factors believed to have potentially potent biological effects on the host. The virulence of staphylococci has been related to toxins such as hemotoxin, epidermolytic (exfoliative) toxin, lethal toxins, leukocidin and enterotoxins (superantigens). Among the non-toxic metabolites are phosphatase, coagulase, hyaluronidase, deoxyribonuclease, staphylokinase or fibrinolysin, lipase, protease and gelatinase.

Infection

Skin disease due to *S. aureus* is the most common of all bacterial infections. It has been estimated that 1.5 million cases of furunculosis occur yearly in the USA.

Folliculitis, furuncles and carbuncles

The most superficial of all staphylococcal skin infections is folliculitis, which is manifested by minute, erythematous, follicular pustules without involvement of the surrounding skin (Figure 2.5). Chronic recurrent folliculitis of the beard area is

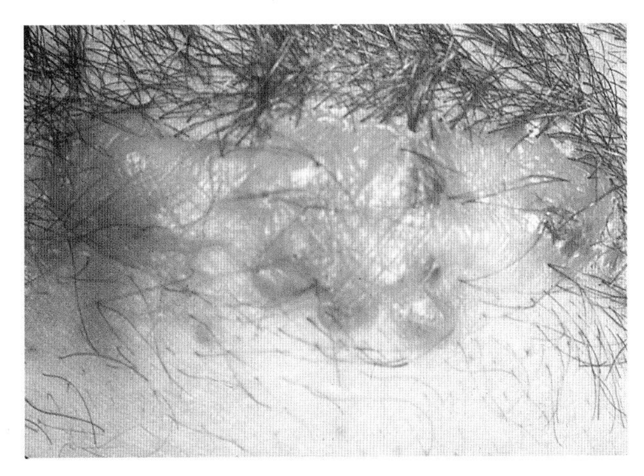

Figure 2.5 Folliculitis of the neck due to staphylococci

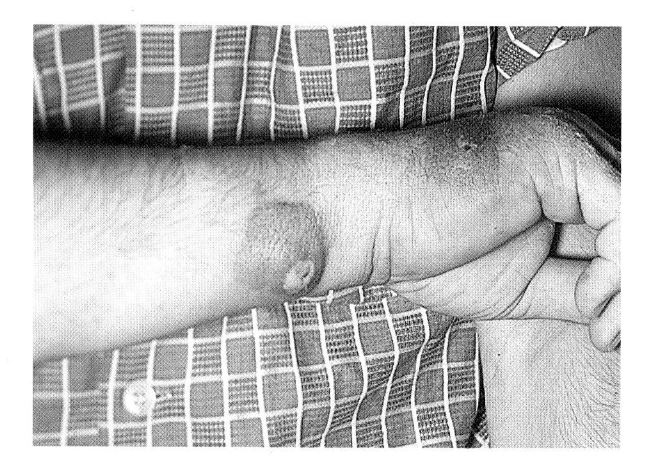

Figure 2.6 A furuncle on the wrist

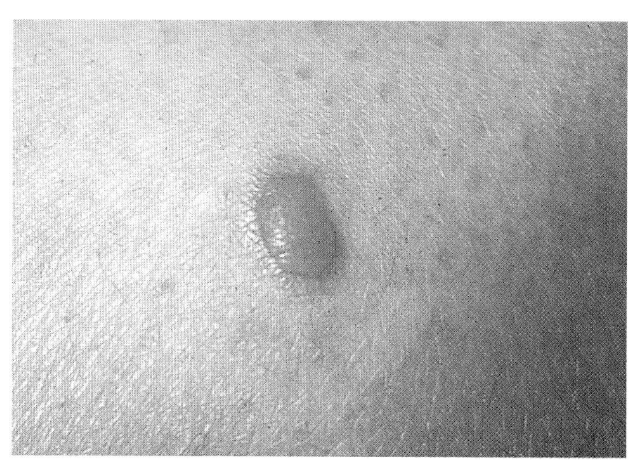

Figure 2.7 Bullous impetigo due to staphylococci

termed sycosis barbae. A furuncle is a deep form of folliculitis with involvement of the subcutaneous tissue (Figure 2.6). On the neck and upper back, multiple hair follicles may be involved, resulting in a carbuncle, which is a large, indurated, painful lesion with multiple draining sites.

Impetigo

There are two classical forms of impetigo: bullous and non-bullous. The bullous type is always induced by *S. aureus*. Flaccid transparent bullae develop on the skin of the face, buttocks, trunk, perineum and extremities. Lesions of bullous impetigo may be considered manifestations of a localized 'staphylococcal scalded skin syndrome' (SSSS; see below) due to exotoxin and, therefore, develop on intact skin (Figure 2.7). Rupture of bullae leaves a narrow rim of scales at the edge of a shallow moist erosion. Non-bullous impetigo typically begins on skin of the face or extremities which has been traumatized. As the appearance of a thick honey-colored crust is virtually pathognomonic of impetigo, cultures are generally not necessary before initiating treatment. Group A β-hemolytic streptococci and *S. aureus* are found together or alone in common impetigo.

Staphylococcal scalded skin syndrome (SSSS)

This syndrome has also been called Ritter's disease or the malignant form of pemphigus neonatorum. Classically, there is diffuse exfoliative disease as a result of systemic toxemia. Epidermolytic toxin is produced by dermatopathic strains of *S. aureus* belonging primarily, but not exclusively, to phage group II, including types 55, 71, 3A, 3B and 3C. Most often, a purulent infection of the upper respiratory tract is present, but any purulent infection, such as omphalitis, conjunctivitis or impetigo, may play this role.

Epidermolytic toxin-producing strains cleave the epidermis longitudinally. Cytological smears of cells on the surface of denuded areas show cells of the stratum granulosum (Tzanck smear). Cultures of material from intact bullae and other affected skin are generally sterile as the staphylococcal infection may lie at a distant focal site. The nose, throat, conjunctiva or any impetiginous area should be swabbed for bacterial culture.

Staphylococcus epidermidis

These gram-positive non-motile cocci (0.5–1.0 μm) occur singly or in pairs, or in irregular clusters that morphologically resemble those of *S. aureus*. *Staphylococcus epidermidis* is coagulase-negative and is thus easily differentiated from *S. aureus*.

Colony morphology

Colonies of *S. epidermidis* are usually circular and convex, with a smooth or slightly granular surface.

They are often white but, occasionally, yellow or orange colonies are seen that may be mistaken for *S. aureus* (Figure 2.8).

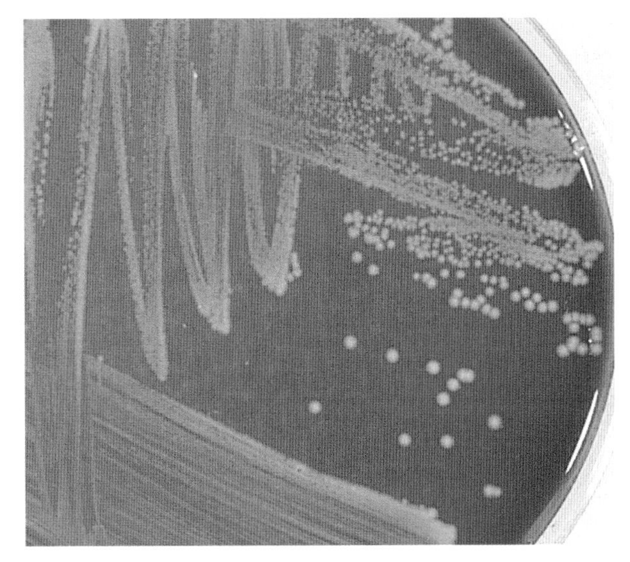

Figure 2.8 Colonies of *Staphylococcus epidermidis* on blood agar

Infection

Penetration of coagulase-negative staphylococci such as *S. epidermidis* beyond the stratum corneum may lead to serious infections. Their involvement in certain infections of the heart valve, bloodstream and urinary tract (*S. saprophyticus*) should be considered. As they are resistant to antibiotics, it is imperative that coagulase-negative staphylococci be tested for susceptibility prior to the administration of antibiotics.

Bibliography

Aly R. The pathogenic staphylococci. *Semin Dermatol* 1990;9:292–9

Aly R, Levit S. The changing spectrum of streptococcal and staphylococcal disease. In: Dahl M, Lynch P, eds. *Current Opinions in Dermatology*. Philadelphia: Current Science, 1993:290–5

Feingold DS. The changing spectrum of streptococcal and staphylococcal infections. In: Aly R, Beutner K, Maibach HI, eds. *Cutaneous Infection and Therapy*. New York: Marcel Dekker, 1997:15–24, 29–33

Maibach HI, Aly R. Bacterial infections of the skin. In: Moschella SL, Hurley HJ, eds. *Dermatology, 3rd edn*. Philadelphia: WB Saunders, 1992:710–50

Noble WE, White MI. Staphylococcal skin infection in man. In: Easmon C, Adlam C, eds. *Staphylococci and Staphylococcal Infections*. London: Academic Press, 1983:165–91

Chapter 3 Streptococci

Streptococci are gram-positive catalase-negative cocci usually found in chains. Their ability to cause hemolysis of erythrocytes to various degrees aids in their classification: β-hemolytic streptococci are able to produce complete lysis or clearing of the red cells surrounding the colony, and is often more pronounced under anaerobic conditions; α-hemolytic streptococci (viridans streptococci) produce a greenish discoloration due to incomplete erythrocyte lysis; and non-hemolytic colonies have been termed γ-hemolytic.

The β-hemolytic streptococci are further subdivided into serological groups on the basis of group-specific antigen found in the cell wall termed C carbohydrates. Group A β-hemolytic streptococci are responsible for 90–95% of streptococcal infections in humans. These hemolytic streptococci are also referred to as *Streptococcus pyogenes*. In group A streptococci, M protein is not only specific for the type, but is essential for virulence. Thus, immunity to M protein is equivalent to having immunity to that strain of streptococcus.

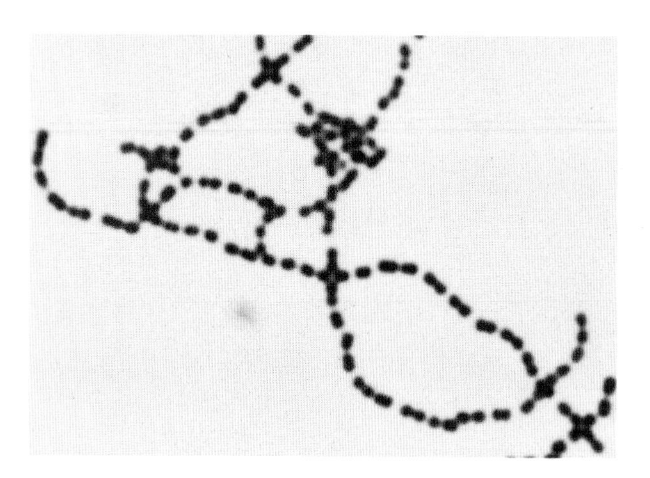

Figure 3.1 Streptococci are usually seen in chains

Streptococcus pyogenes is rarely seen on normal human skin in hygienic conditions, although it may be carried in the throat and nares of approximately 10% of the general population. The paucity of *S. pyogenes* on the skin is attributed to the presence of skin lipids, which are lethal *in vitro* for streptococci.

Morphology

Streptococci are spheres 0.5–1.0 μm in diameter, found in pairs and in characteristic short-to-long chains (Figure 3.1), although chain formation may be lost when they are subcultured repeatedly on solid media. These organisms are non-motile gram-positive cocci. In actively spreading lesions within tissues, diplococcal and individual coccal forms are commonly seen whereas, in purulent exudates from walled-off lesions and in broth culture, streptococci are most commonly found in chains.

Specimen collection

After removal of the roof or crust of the pustule or vesicle with a sterile Bard–Parker number 7 blade, sterile swabs are used to collect the bacteria. Wound cultures are treated similarly. Exudate, if present, should be touched with the swab. Throat swabs are taken by vigorous swabbing of the tonsil and posterior pharynx areas. Taking care to avoid touching the tongue or cheeks with the swab reduces contamination.

Direct microscopic examination of smears from throat swabs is of limited value in the specific detection of β-hemolytic streptococci as other streptococci normally colonize the throat.

Direct microscopy of pus from a closed streptococcal pustule or vesicle may reveal gram-positive cocci. This is mainly seen in cases of streptococcal impetigo. Culture is required to determine whether the cocci are staphylococci or streptococci.

It is difficult to demonstrate organisms in either cellulitis or erysipelas. Culture is therefore performed by injecting preservative-free physiological saline into the involved skin and culturing the withdrawn fluid as soon as possible, or by skin biopsy. However, despite meticulous technique, it is unusual to culture streptococci from these lesions.

Swabs should be inoculated on blood agar plates immediately after collection, although delay in delivering the swabs to the laboratory cannot be avoided in certain situations. Group A streptococci survive well if the swab is quickly dried and kept dry during shipment to the laboratory by inserting the swabs either in a screw-cap tube or an aluminum-foil packet containing indicator silica gel. Alternatively, swabs may be rolled and scrubbed onto a piece of sterile filter paper, allowed to air-dry for 30–40 min, then folded within the carrier paper before shipping. These methods limit the multiplication of organisms, maintain viability of the streptococcal population and reduce the deleterious effects of other bacterial flora.

Culture and identification

Streptococci grow readily on enriched artificial media. Sheep's blood agar is recommended for primary isolation. The pathogenic forms grow best at 37° C. Most strains are facultative to obligate anaerobes. On blood agar plates, group A streptococci form small grayish colonies visible within 18–24 h. A few stabs with the loop through inoculated areas of the agar permit observation of hemolysis that may not have occurred on the surface (Figure 3.2).

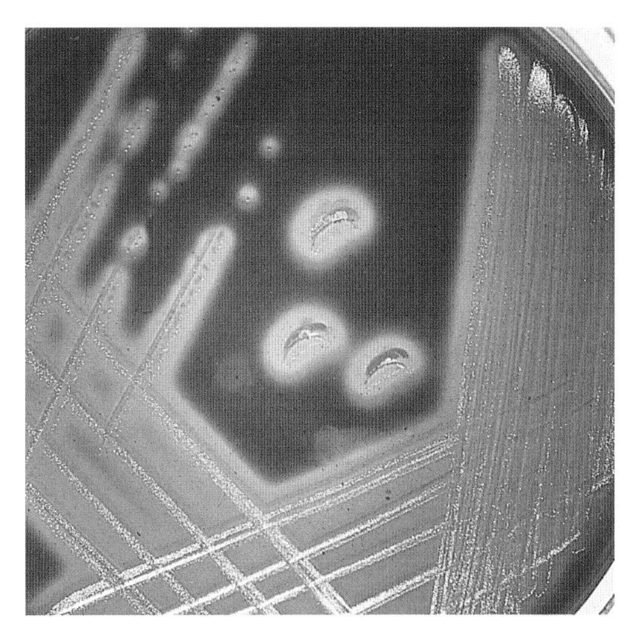

Figure 3.2 Group A streptococci showing β hemolysis surrounding the colonies

Hemolytic streptococci may be mistaken for hemolytic staphylococci, but may be differentiated by the catalase test. Hydrogen peroxide (0.5 ml of a 3% solution) is added to colonies on plain (without blood) agar slants or plates. The production of gas bubbles is seen with staphylococci, but not streptococci.

A convenient presumptive test for differentiating group A from other β-hemolytic streptococci is the bacitracin disk test. The test is based on the selective inhibition of group A streptococci on blood agar containing bacitracin 0.04 U (Figure 3.3). Some group B streptococci may also be bacitracin-sensitive.

When the culture is contaminated with other bacteria, sheep's blood agar with 1 μg / ml of crystal violet stain is superior to plain blood agar as an isolation medium for β-hemolytic streptococci.

Figure 3.3 Bacitracin test showing inhibition of group A hemolytic streptococci around the bacitracin disk

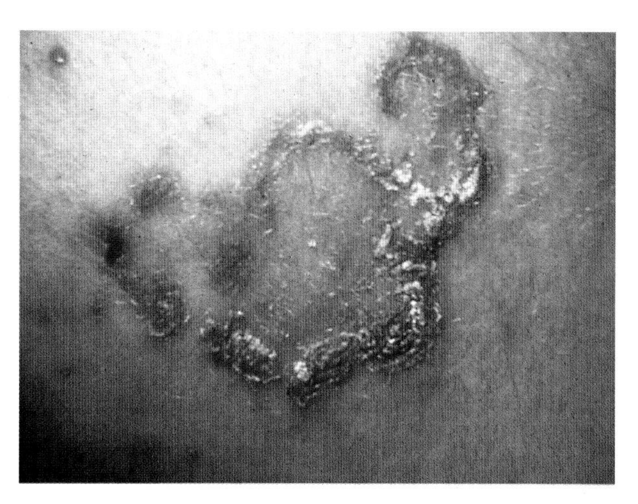

Figure 3.4 Superficial (or common) impetigo. Courtesy of Dr Timothy Berger, San Francisco, CA

Serological tests

Antibodies to some of the extracellular growth products of streptococci may be used in diagnosis. The antistreptolysin O (ASO) titer, which peaks at 2–4 weeks after acute infection, and antinicotinamide-adenine dinucleotide (NAD)ase titer, which peaks at 6–8 weeks after acute infection, are more commonly elevated after pharyngeal than after skin infections. In contrast, antihyaluronidase is elevated after skin infections, and antideoxyribonuclease (DNase) B rises after both pharyngeal and skin infections. Titers during late sequelae (acute rheumatic fever and acute glomerulonephritis) reflect the site of primary infection. Although it is not as well known as the ASO test, the antiDNase B test appears to be superior because high-titer antibody is detected following skin and pharyngeal infections, and during late sequelae. Those titers should be interpreted in terms of the age of the patient and geographical locale.

Erythrogenic toxin

Erythrogenic toxin is responsible for the rash seen in scarlet fever. Strains of group A streptococci that produce the toxin are considered lysogenic, such as the toxigenic strains of *Corynebacterium diphtheriae*. The mode of action of the toxin is not clear but, when injected into the skin of a susceptible subject, it triggers a localized erythematous reaction which reaches maximum at around 24 h; this is termed the Dick test. The erythrogenic effect is neutralized when the patient's serum contains antitoxin and the skin test becomes negative. A positive Dick test indicates an absence of circulating antitoxin and a state of susceptibility to scarlet fever. An injection of homologous antitoxin intradermally at the peak of scarlet fever causes local blanching of the rash or the Schultz–Charlton reaction.

Other common extracellular products are NADase (leukotoxic), streptokinase, streptodornase, hyaluronidase and protease.

Infection

Group A streptococci are associated with several forms of skin and soft-tissue infections. In patients with streptococcal skin infections, lesions are typically vesicular or pustular in the early stages and later become covered with a thickened and honey-colored crust. Pus formed as a result of streptococcal infection may be somewhat thinner than that found in a staphylococcal abscess. Streptococcal skin infections are an important cause of acute glomerulonephritis.

Impetigo

This infection (Figure 3.4) first appears as a vesicle, rapidly becomes pustular and is then covered by a thickened crust. Vesicles and fresh pustules yield more pure cultures of streptococci than older crusted lesions, which yield mixed cultures of streptococci and staphylococci. Lymphatic involvement is common, with transient adenitis followed by regional lymphadenopathy. In many patients, the primary organism is *Staphylococcus aureus*, which may or may not be secondarily contaminated with streptococci.

Scarlet fever

An erythematous rash (Figure 3.5) may follow a streptococcal sore throat if the infecting streptococci produce erythrogenic toxin and the patient has no antitoxic immunity. Peritonsillar abscesses (quinsy) or Ludwig's angina may block the airways by massive swelling of the floor of the mouth. Acute hemorrhagic glomerulonephritis and acute rheumatic fever are the most serious complications of scarlet fever.

Cellulitis

This infection (Figure 3.6) is a complication of a wound or ulcer, but it may develop in previously normal skin, especially in the presence of edema of lymphatic, renal or hypostatic origin. A diffuse brawny inflammation of the skin and subcutaneous tissue is seen. A circumscribed indurated lesion that pits on pressure then develops, and is dusky red, hot and tender with poorly defined borders. There is no absolute distinction between streptococcal cellulitis and erysipelas, although the latter is more superficial with sharper margins.

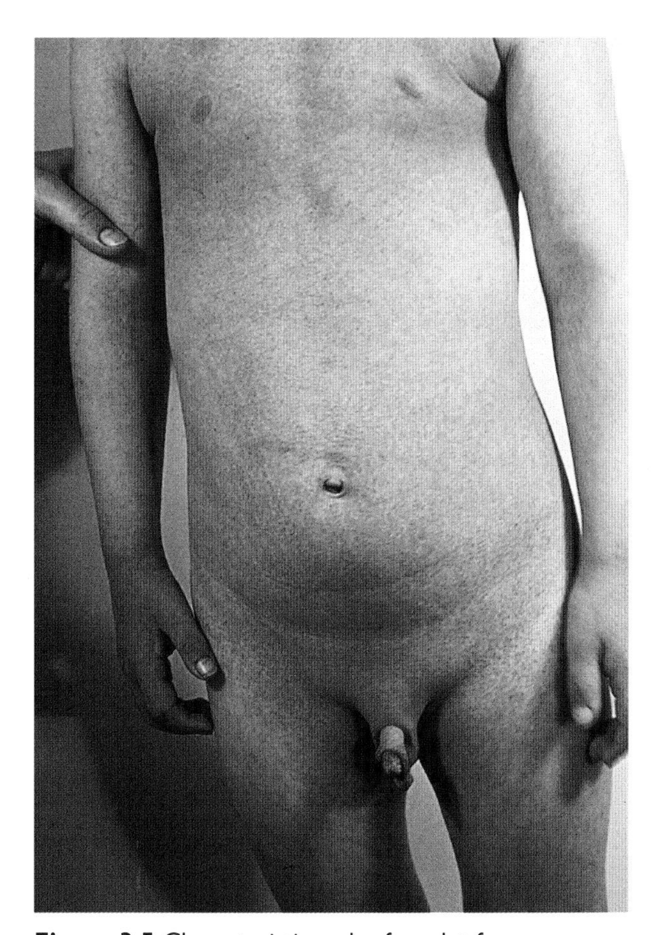

Figure 3.5 Characteristic rash of scarlet fever

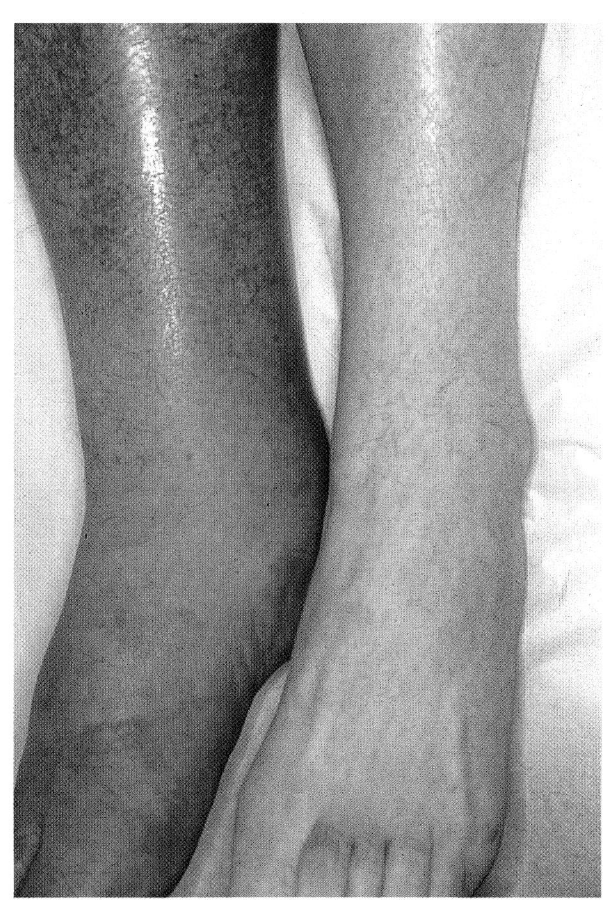

Figure 3.6 Cellulitis involving one foot and leg

Erysipelas

Although this disease is contagious from one person to another, it does not result in explosive epidemics such as seen with scarlet fever. In fact, erysipelas is infrequently seen nowadays. The microorganism is found in the edema fluid of the advancing margin and characteristically spreads within the subepidermal tissues. Bacteremia arises more often in patients with erysipelas and deeper forms of infection than in those with superficial pyoderma.

Bibliography

Feingold DS. The changing spectrum of streptococcal and staphylococcal infections. In: Aly R, Beutner K, Maibach HI, eds. *Cutaneous Infection and Therapy.* New York: Marcel Dekker, 1997:15–24, 29–33

Hoge CW, Schwartz B, Talkington DF, *et al.* The changing epidemiology of invasive group A streptococcal infections and the emergence of streptococcal toxic shock syndrome. *J Am Med Assoc* 1993;269:384–9

Maibach HI, Aly R. Bacterial infections of the skin. In: Moschella SL, Hurley HJ, eds. *Dermatology, 3rd edn.* Philadelphia: WB Saunders, 1992:710–50

Stevens DL. Invasive group A staphylococcal infections. *Clin Infect Dis* 1992;14:2–11

Chapter 4 Corynebacteria and Propionibacteria

CORYNEBACTERIA

Corynebacteria are gram- and catalase-positive pleomorphic rods containing metachromatic (volutin) granules demonstrable by Albert's stain (methylene blue). Dermatologically speaking, the three important species are *Corynebacterium diphtheriae*, *C. minutissimum* and *C. tenuis*. *Corynebacterium diphtheriae* is primarily an inhabitant of the nasopharynx that establishes itself in damaged skin. The term 'diphtheroid' refers to a wide range of gram-positive pleomorphic rods of the genus *Corynebacterium* other than *C. diphtheriae*.

Erythrasma

Corynebacterium minutissimum is the etiological agent of erythrasma. The bacterium is part of the resident flora, and is readily recovered from moist areas of the skin such as the axillae, groin and toewebs. The organism produces porphyrin fluorescence in certain media, and is responsible for the coral-red fluorescence seen with clinical lesions.

Direct examination

Because of its small size, the organism is difficult to observe in KOH preparations, but is readily demonstrable by Gram's stain of the stratum corneum collected by tackified adhesive, or by the cyanoacrylate adhesive slide technique. When positive, the scales reveal gram-positive, pleomorphic, often thin rods with or without granules (Figure 4.1). Gram's stain of a smear from a colony shows gram-positive pleomorphic rods (Figure 4.2).

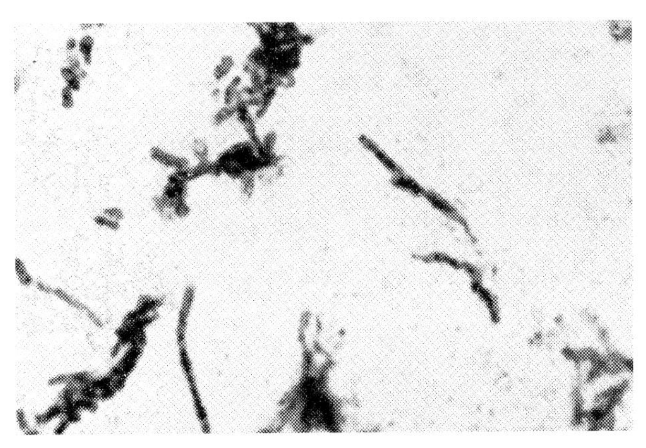

Figure 4.1 Corynebacteria in the horny layer of skin from a patient with erythrasma of the groin (Gram's). Reproduced with permission, from Sarkany & Taplin, *J Invest Dermatol* 1961;37:283

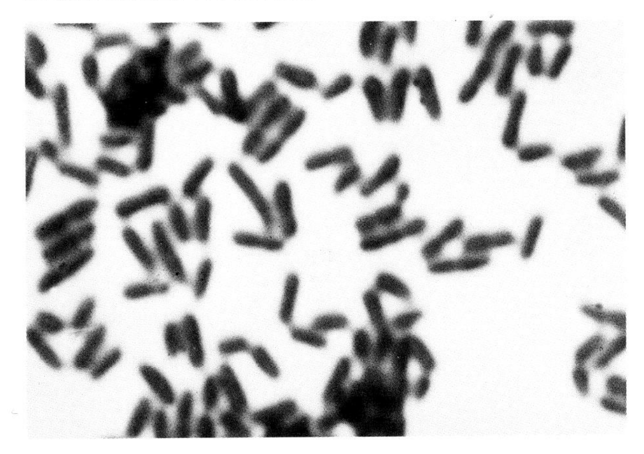

Figure 4.2 *Corynebacterium minutissimum* in a smear from culture (Gram's; ×900)

Culture

When grown on standard culture media with added 20% fetal bovine serum, red colonies of *C. minutissimum* are observed within 2–6 days. The colonies and the medium surrounding them show coral-pink fluorescence when examined under Wood's light. The porphyrin responsible for the fluorescence is water-soluble and diffused through the agar around the colonies.

Infection

The microorganism colonizes the pubis, toeweb, groin (Figure 4.3), intergluteal cleft, axilla and inframammary folds. It produces clinical disease infrequently. Most lesions are asymptomatic, but may be mildly symptomatic with burning and itching. The patches are irregularly shaped, dry and scaly, initially pink and later turning brown. The less common generalized form of erythrasma is predominantly seen in obese subjects.

Corynebacterium minutissimum may colonize toewebs infected with fungi. Erythrasma at this site may closely resemble tinea pedis (athlete's foot), and is characterized by scaling, fissuring and slight maceration. The bacteria are not seen in routine KOH preparations because of their small size.

Trichomycosis

Three types of diphtheroids have been associated with trichomycosis: One resembles *C. minutissimum*; another is lipophilic; and the third is *C. tenuis*. These bacteria contain keratolytic enzymes, as evidenced by electron micrographs of affected hairs showing destruction of the cuticular keratin.

Direct examination

Infected hairs are placed on a slide in a drop of 10% KOH under a coverslip. The preparation may be heated gently for clearing. The nodules on the hair (Figure 4.4) are composed of short bacillary forms ≤1 μm in diameter which, when crushed, are seen to be embedded in mucilaginous material.

Culture

These microorganisms are difficult to culture as the bacteria associated with the lesions are concentrated in granules and encrustations that resist mechanical dispersion. Therefore, infected hairs are partially degermed by immersion in 70% alcohol for 1 s, then placed for 18–24 h in saline buffered with pH 7.2 and containing 1% Triton X-100. The granules are then removed, macerated with dissecting needles and cultured on brain–heart infusion agar containing 1% Tween 80.

Infection

The infection involves the hair shafts in the axillary and pubic regions, and is characterized by the development of nodules that vary in consistency and color. The nodules may be yellow, red or black, and appear along the hair shaft as either discrete and scattered forms or as a continuous sheath. The infection does not extend into the fol-

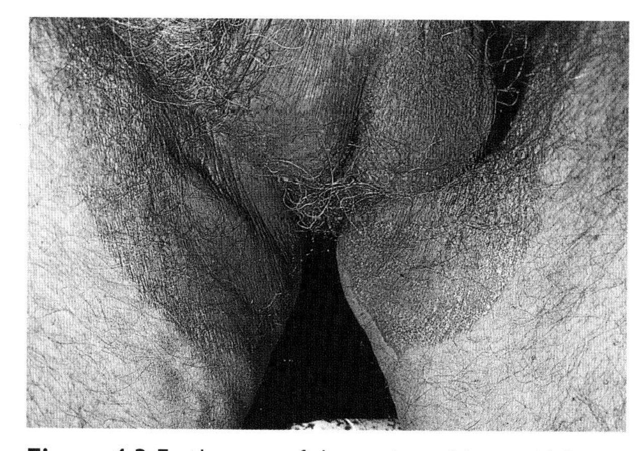

Figure 4.3 Erythrasma of the groin and inner thighs

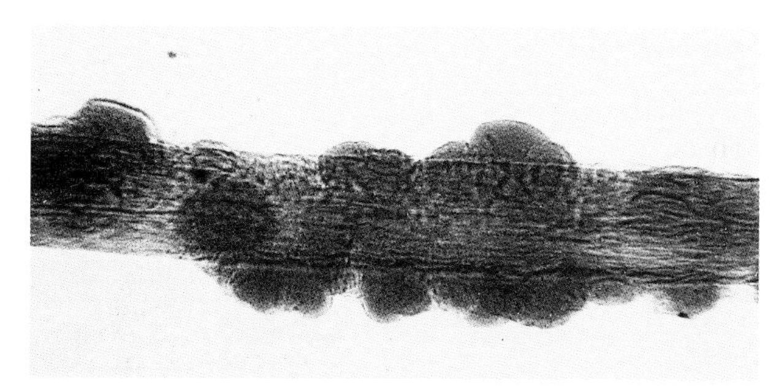

Figure 4.4 Trichomycosis of axilla hair. Note nodules around the hair (KOH)

licle or involve the surrounding skin. Examination under Wood's light is helpful as the infected hairs are then visible as a bright blue to pink or white fluorescence.

Pitted keratolysis

The gram-positive coryneform bacteria *Micrococcus sedentarius* and *Dermatophilus congolensis* have been isolated from pitted keratolytic lesions. The microorganisms grow both aerobically and anaerobically. The filamentous form is dominant with the latter and the diphtheroid form is favored with aerobic culture. *Actinomycetes* was once implicated as the causative agent.

Direct smear

Gram's staining of ground stratum corneum reveals both filamentous and coccal forms of the microorganisms (Figure 4.5). Gram-stained biopsy shows the presence of microorganisms in the floor and walls of the pits (Figure 4.6).

Infection

Pitted keratolysis is a superficial infection of the plantar surface presenting a punched-out appearance. Filamentous microorganisms infiltrating the stratum corneum are confined entirely to the area of the erosion. The pits, produced by a lytic process that spreads peripherally, may coalesce into irregularly shaped patches of superficial erosion. The areas most often infected are the heel, ball and volar pads of the foot, and the toes. Humidity and high temperature are frequently aggravating factors.

PROPIONIBACTERIA

Three species of *Propionibacterium* are common inhabitants of human skin: *P. acnes* and *P. granulosum*, found mainly in sebaceous areas, and *P. avidum*, seen mostly in the axillae. *Propionibacterium acnes* is eight times as frequently isolated in acne lesions as is *P. granulosum* and is probably involved in the pathogenesis of acne.

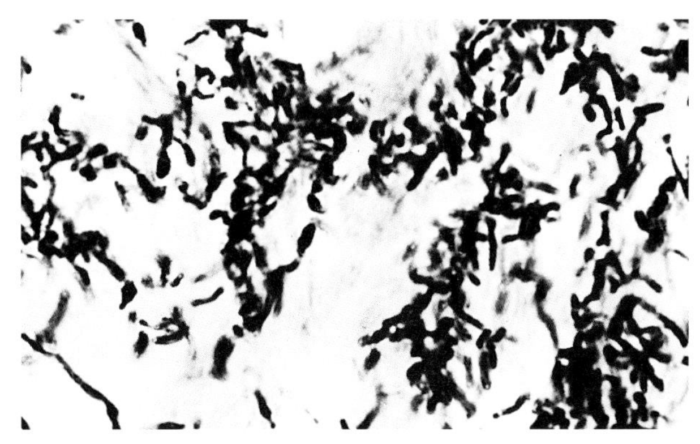

Figure 4.5 Ground stratum corneum shows both filamentous and coccal bacterial forms in pitted keratolysis (Gram's)

Figure 4.6 In pitted keratolysis, bacteria are seen at the base of the pit, which is confined to the stratum corneum (methenamine–silver). Reproduced with permission, from Zaias *et al.*, *Arch Dermatol* 1965;92:151

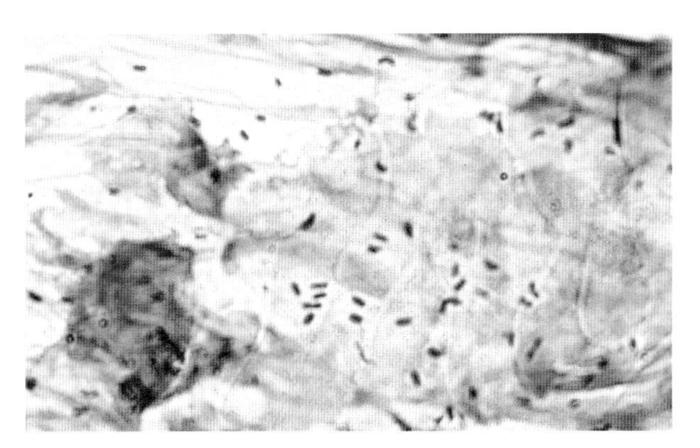

Figure 4.7 Smear from a pustule in acne (Gram's)

These bacteria resemble coryneforms, being gram-positive pleomorphic rods that vary in length from 0.5–2.0 μm. However, in contrast to coryneforms, they are strictly anaerobic, but are able to tolerate microaerophilic conditions.

Specimen collection and culture

The acne pustules are cleansed with 70% alcohol, and specimens taken with a number 11 scalpel blade. The pus is spread thinly onto a clean glass slide for Gram's staining. Numerous gram-positive pleomorphic rods as well as gram-positive cocci are seen ingested in white blood cells (Figure 4.7). Brewer's thioglycollate with 1.5% agar, trypticase soy agar or brain–heart infusion agar with 1% glucose are satisfactory as culture media.

Propionibacterium acnes grows anaerobically in a GasPak® system (BBL) at 37°C for 3–5 days. The colonies vary in color from white to salmon-pink and from 1–3 mm in diameter. The colonies are round, raised, smooth and glistening. The colony characteristics of *P. acnes* and *P. granulosum* are identifiable on modified Marsal and Kelsey casein–yeast lactate glucose agar (Figure 4.8).

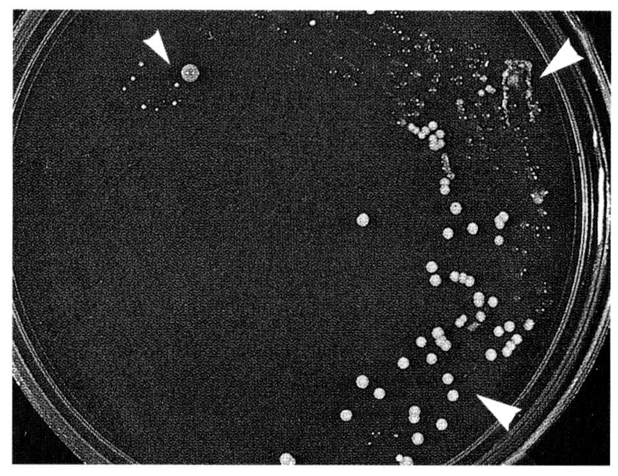

Figure 4.8 Colonies of anaerobic propionibacteria grown on a casein–yeast lactate glucose agar, including a small number of large reddish-brown colonies of *Propionibacterium granulosum* (small arrow) and numerous medium-sized pinkish colonies of *P. acnes* (large arrows). Courtesy of K.J. McGinley, Philadelphia, PA

Acne vulgaris

Acne (Figure 4.9) is a disease of the pilosebaceous follicles. The role of bacteria in acne has received considerable attention due to the successful inhibition of the disease by antimicrobial agents. *Propionibacterium acnes* hydrolyzes triglycerides secreted by the sebaceous gland into diglycerides, monoglycerides and free fatty acids. As fatty acids are irritants when injected into the human skin as well as comedogenic, free fatty acids have been implicated, but not confirmed, as pathogenic factors in acne. Gram-positive cocci and *Pityrosporum* are also present in the sebaceous glands.

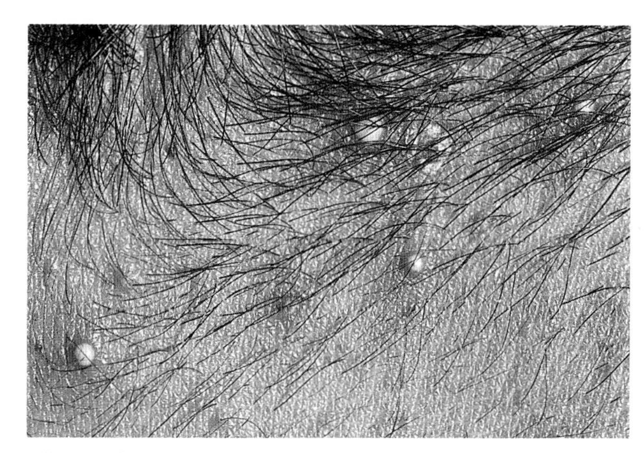

Figure 4.9 Acne vulgaris lesions on the back of the neck

Acne is now considered to be a multifactorial disease. The principal pathogenic events in acne are: (1) abnormal follicular keratinization leading to plugging of the follicle; (2) increased sebum production behind the follicular plug; (3) proliferation of *P. acnes* in the sebum; and (4) inflammation.

Bibliography

Golledge CL, Phillips G. *Corynebacterium minutissimum* infections. *J Infect* 1991;23:73–6

Noble WC. Coryneform bacteria of human skin. In: Aly R, Maibach HI, eds. *Atlas of Infection of the Skin.* London: Churchill Livingstone, 1998

Stanton RL, Schwartz RA, Aly R. Pitted keratolysis: A clinicopathologic review. *J Am Podiatr Assoc* 1982; 72:436–9

Chapter 5 Gram-negative bacteria

From a clinical cutaneous viewpoint, there are four groups of gram-negative bacilli: *Pseudomonas* spp; *Proteus* spp; coliform-like organisms (heterogeneous groups such as *Enterobacter*, *Alcaligenes*, *Escherichia* and *Serratia*); and *Acinetobacter*. Microscopically, these bacilli are not differentiable from one another with Gram's stain.

Pseudomonas aeruginosa

This microorganism is most frequently implicated in human infections. Normal dry human skin does not support colonization of this organism but, in situations where the stratum corneum is abnormal, *P. aeruginosa* may flourish and induce several specific dermatoses. Patients may become susceptible to *P. aeruginosa* after prolonged treatment with immunosuppressive agents, corticosteroids, antimetabolites and antibiotics.

Morphology

Pseudomonas aeruginosa is a gram-negative bacillus, measuring 1–2 μm in length, seen singly, in pairs or in short chains. The organism is motile and possesses 1–3 polar flagella.

Culture

The bacterium is aerobic and grows on most laboratory media at an optimal growth temperature of 37°C. Overnight incubation on nutrient agar results in large, soft, smooth, grayish, spreading colonies which may become confluent and cover the entire culture surface. *Pseudomonas aeruginosa* elaborates water-soluble pigments which diffuse into *Pseudomonas* isolation medium (Difco, Detroit, MI; Figure 5.1), but are not seen on blood agar.

The presence of this organism may be suspected by its characteristic aromatic grape-like odor. *Pseudomonas* agar (Difco) encourages pyocyanin synthesis. As *P. aeruginosa* is a non-lactose fermenter as well as oxidase- and catalase-positive, it uses glucose oxidatively in oxidation–fermentation (O–F) medium. For each isolate, two tubes of glucose O–F medium are inoculated with a light stab from an overnight culture, one of which is overlaid with 6 mm of sterile stiff petroleum or mineral oil. The tubes are incubated at 37°C for several days

Figure 5.1 *Pseudomonas aeruginosa* produces a blue pigment that diffuses into the culture medium

and examined daily. A change from blue-green to yellow constitutes a positive reaction.

Infection

The pathogenic nature of this microorganism in a healthy subject is limited, usually presenting as hot-tub folliculitis, toeweb infection, green nail syndrome or otitis externa.

Hot-tub folliculitis

In this infection, the lesion appears 2–3 days after the episode of immersion in a hot tub. The vesicle on an erythematous base involves the trunk, buttocks, legs and arms (Figure 5.2). The microorganism is readily cultured from the pus and the water from the tub. Gram's staining of pus reveals gram-negative rods and neutrophils.

Toeweb infection

The most common site of *P. aeruginosa* infection in the general population is the toeweb. Typical hyperkeratotic, white and intertriginous lesions may show a greenish discoloration due to the pigments elaborated by the bacterium. Wood's light examination reveals aqua-green fluorescence. *Pseudomonas cepacia* may colonize moist toewebs, and has been reported in urinary tract infections, wound sepsis, endocarditis and septicemia.

Green nail syndrome

This refers to the clinical appearance of greenish discoloration of the nail plate, which is commonly associated with paronychia and the presence of *P. aeruginosa*. In chronic paronychia, there is often concomitant *Candida albicans* colonization or infection of the nail fold. Black-colored paronychia is associated with *Proteus* infection.

Other clinical disorders seen less frequently are malignant external otitis, ecthyma gangrenosum and *Pseudomonas* septicemia. Early recognition of these infections allows specific lifesaving treatment.

Enterobacteriaceae

Members of the Enterobacteriaceae family (*Escherichia coli*, and *Klebsiella*, *Enterobacter* and *Proteus* spp) have limited potential for invading the skin. *Proteus vulgaris* and *P. mirabilis* are the exceptions, as they produce proteolytic enzymes which may play a role in foul-smelling wounds, ulcers, cystic acne and toeweb lesions.

Culture

Proteus is a gram-negative, motile, aerobic bacillus which does not ferment lactose, but rapidly liquefies gelatin or decomposes urea with liberation of ammonia. When grown on blood agar or trypticase soy agar, these bacteria spread in waves over the

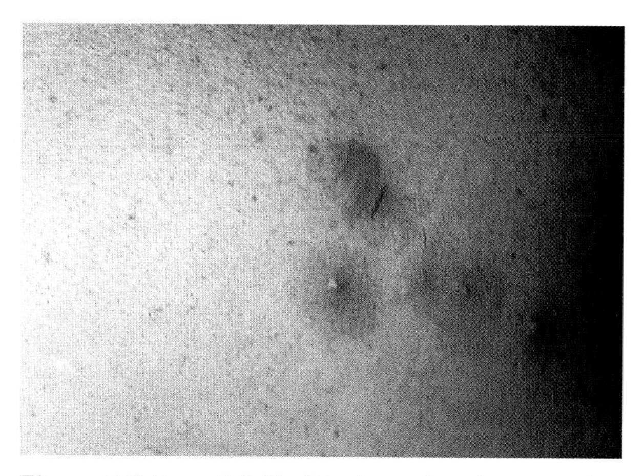

Figure 5.2 Hot-tub folliculitis due to *Pseudomonas* infection. Courtesy of Dr. Timothy Berger, San Francisco, CA

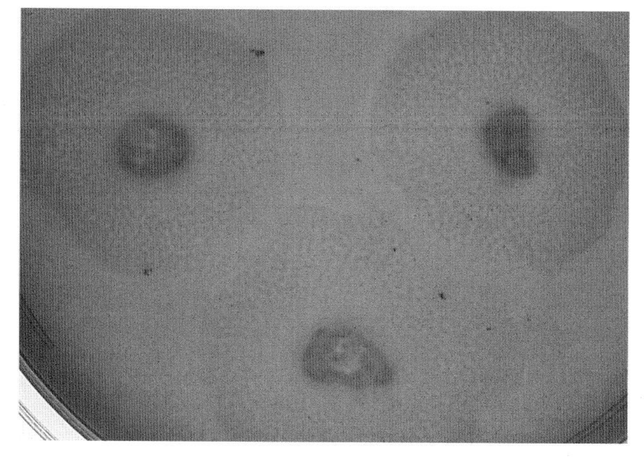

Figure 5.3 *Proteus* grown on sheep's blood agar. Note the spreading of the colony

Figure 5.4 Colonies of *E. coli* grown on EMB agar are characterized by a blue center and a greenish metallic sheen

Figure 5.5 Colonies of *E. aerogenes* grown on EMB agar

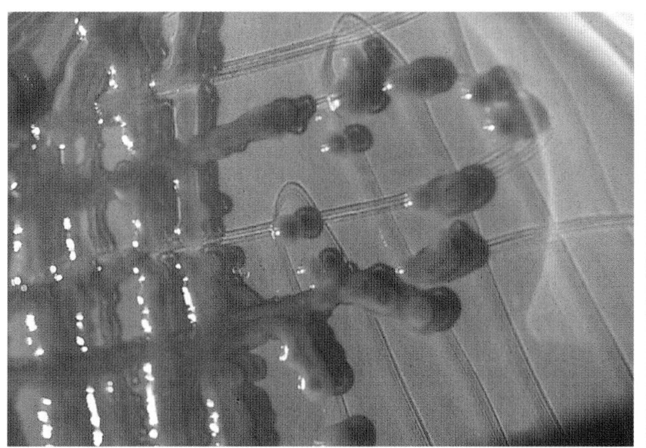

Figure 5.6 Colonies of *K. pneumoniae* grown on EMB agar (left) and sheep's blood agar (right)

surface of solid media (Figure 5.3). Its presence is detected by a characteristic foul smell. On eosin–methylene blue (EMB) agar, *Proteus* produces translucent and colorless colonies whereas, in contrast, *E. coli* colonies usually present with a dark-blue center on EMB with a greenish metallic sheen (Figure 5.4). On blood agar, *E. coli* forms non-hemolytic colonies.

Colonies of *Enterobacter aerogenes* are frequently larger than those of *E. coli* and may coalesce (Figure 5.5). Colony centers are usually brown or purple and not as dark as with *E. coli*, and a metallic sheen is rare.

Klebsiella spp are non-motile, encapsulated, short rods which, when cultured on EMB and blood agar, produce large mucoid colonies that have a tendency to coalesce (Figure 5.6). Colony growths usually form strings when touched with a needle.

The biochemical reactions of various gram-negative bacteria are shown in Table 5.1.

Infection and specimen collection

Infections caused by this group of gram-negative bacteria have increased in importance as a result of the use of antibiotics effective against gram-positive bacteria.

Folliculitis

This syndrome is usually associated with the older-

Table 5.1 Biochemical reactions of gram-negative bacteria commonly found on the skin

	Pseudomonas aeruginosa	*Alcaligenes* spp	*Enterobacter* spp	*Escherichia coli*	*Proteus vulgaris*	*Klebsiella* spp
Iron butt	–	–	+gas	+gas	+gas	+gas
slant			+	+	+	+
H_2S			–	–	+	–
Phenylalanine deamination	–	–	–	–	+	–
Urease	–	–	–	–	+	+
Indole	–	–	–	+	+	–
Methyl red	–	–	–	+	+	–
Voges–Proskauer	–	–	+	–	–	+
Citrate	+	+	+	–	±	+
Lactose	–	–	+	+	–	+
Gelatin	+	–	–	–	+	–
Pyocyanin agar	+	–	–	–	–	–
Motility	+	+	+	±	+	–
Dextrose	+	–	+	+	+	+
EMB medium	colorless	colorless	pink mucoid	green metallic	colorless	blue

age acne group who have received several broad-spectrum antibiotics for extended periods of time. Two clinical variants are seen; the superficial pustular type usually occurs around the nose and is associated with *Klebsiella* or *Enterobacter,* whereas the deep nodular and cystic type is associated with *Proteus* infection.

The pustule selected for culture should be fresh and rich in pus rather than an older dried-out lesion. Because gram-negative bacteria are sensitive to desiccation, the pus should be cultured as soon as possible. The microorganisms are rarely found on a direct gram-stained smear and thus require culture. A culture sample should also be taken from the nose and, in cases of true infection (rather than colonization), should demonstrate the same organism in reasonable numbers.

Infection of the sole of the foot

On rare occasions, a vesiculobullous eruption develops on the sole which is often mistaken for a severe dermatophytosis pedis. Some cases begin as a KOH-positive dermatophytosis, but the dermatophyte is slowly replaced by gram-negative bacteria. In others, only gram-negative microorganisms are found. In either instance, the infection is severe, persistent and often disabling. Gram-negative disease should be considered in plantar vesiculobullous eruptions.

Generous quantities of involved stratum corneum are either scraped off, using a number 15 scalpel blade, from the toeweb, for example, or clipped off using an iris scissors, in cases of bullae on the soles. These samples are minced in a drop of saline or water on a glass slide or ground up with a few drops of fluid in a hand-held glass homogenizer. The minced stratum corneum is Gram-stained. Luxuriant numbers of gram-negative rods may be observed. The gram-negative rods are not generally found in white blood cells, but in the stratum corneum.

Bibliography

Agger WA, Mardan A. *Pseudomonas aeruginosa* infection of intact skin. *Clin Infect Dis* 1995;20:302–8

Aly R, Maibach HI. Aerobic flora of intertriginous skin. *Appl Environ Microbiol* 1977;33:97–100

Grimont PAD, Bouvet PJM. Taxonomy of *Acinetobacter*. In: Towner KJ, Bergogne-Berezin E, Fewson CA, eds. *Biology of the Acinetobacter*. New York: Plenum Press, 1991:25–36

Noble WC. Skin infections with gram-negative bacilli. In: Aly R, Beutner K, Maibach HI, eds. *Cutaneous Infection and Therapy*. New York: Marcel Dekker, 1997:29–35

Chapter 6 Other cutaneous bacteria

Neisseria

These gram-negative diplococci comprise paired cells with flattened adjacent walls. The two species of pathogenic importance are *Neisseria meningitidis* and *N. gonorrhoeae*. These bacteria are extremely fastidious, and require enriched media and incubation at an increased carbon-dioxide tension. Gonococci are obligate human parasites not found in other animals.

Direct smear

Pus and secretions are taken from gonococcal lesions of the urethra, cervix, prostate and, occasionally, rectal mucosa. Smears are often more useful when taken during the acute phase of infection than the chronic phase. However, cultures generally yield a higher percentage of positive results.

Direct smears are prepared by rolling the swab over a slide and Gram-staining to reveal the presence of intracellular gram-negative diplococci (Figure 6.1). The appearance of gram-negative intracellular diplococci constitutes a positive presumptive test for gonorrhea. Depending on the phase of infection, these bacteria may also be seen extracellularly.

Skin lesions have been observed especially in women. Often, these appear at the same time as pudendal involvement. Careful unroofing of the skin lesion is performed with a sterile needle. Culture of scrapings from the base of such a lesion has been successful in a small percentages of cases.

Culture

Ideally, specimens from suspected gonococcal infections should be inoculated onto selective media (such as Thayer–Martin) without delay. These media are commercially available as plates and as packages containing a source of carbon dioxide, thus allowing the culture plate to be readily sent through the mail.

The clinical specimen is best streaked onto Thayer–Martin chocolate medium immediately after collection and incubated in an atmosphere of carbon

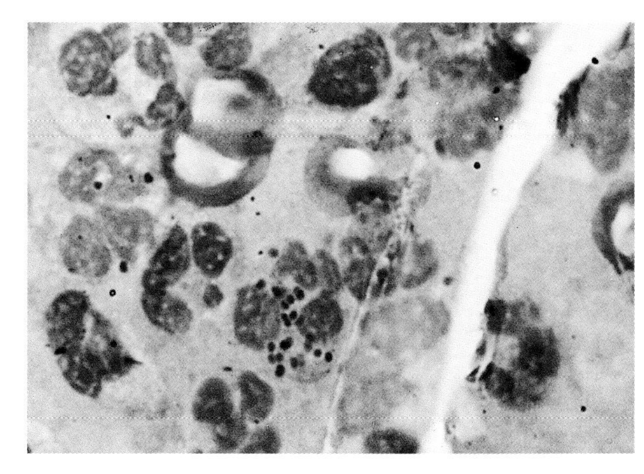

Figure 6.1 Direct smear from the urethra shows diplococcal gonococci within white blood cells

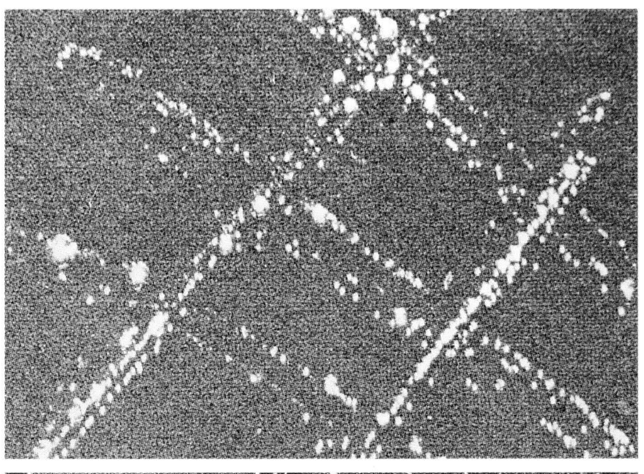

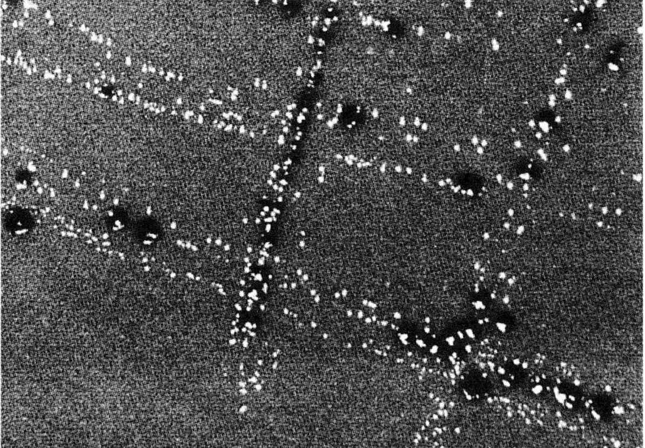

Figure 6.2 Colonies of *N. gonorrhoeae* appear grayish on Thayer–Martin chocolate medium (upper), but turn black after application of redox dye (lower), indicating oxidase-positivity

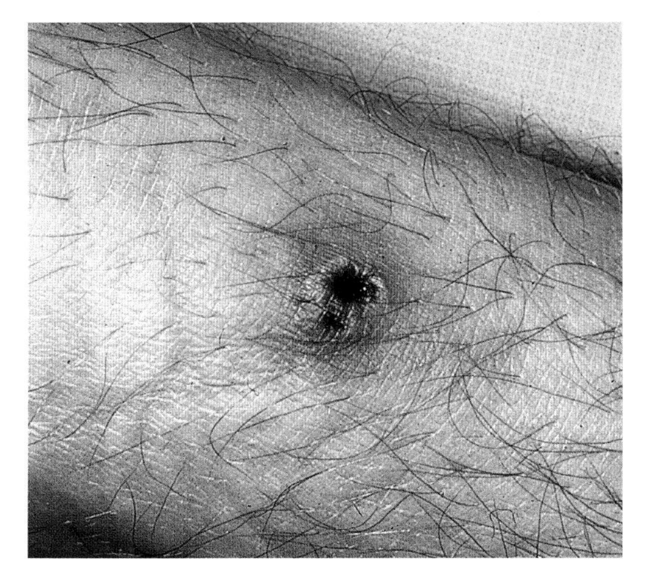

Figure 6.4 Hemorrhagic inflamed pustule of disseminated gonococcal infection

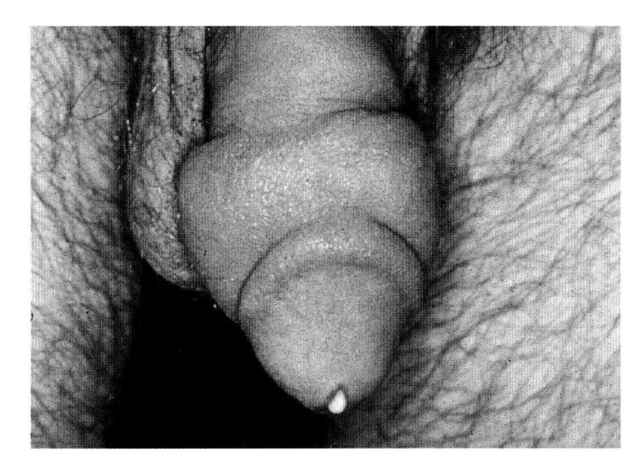

Figure 6.3 Purulent gonorrheal urethral discharge

dioxide (candle jar) at 35–36°C for 24–48h. The colonies are small, translucent, moist and grayish in color. When 0.5% tetramethyl-p-phenylenediamine dihydrochloride (a redox dye) is applied with a wire loop to a suspected colony on the plate, the color of the colony changes from grayish to black (Figure 6.2).

This oxidase test is presumptive identification of the microorganism. Identification of *N. gonorrhoeae* is confirmed by the carbohydrate fermentation test or by the direct fluorescent antibody technique. A specimen from the genitourinary tract inoculated onto Thayer–Martin medium which demonstrates growth of typical oxidase-positive colonies of gram-negative diplococci is sufficient for a presumptive identification of *N. gonorrhoeae*.

Infection

Gonorrhea is primarily a disease of the mucous membrane, and predominantly involves the urethra in men (Figure 6.3), and the urethra, paraurethral glands and cervix in women. Vulval inoculation may occur in children, and conjunctival inoculation has been noted in both children and adults.

Gonococcal bacteremia may lead to skin manifestations. The skin lesions are sparse and transient, and appear in crops on the limbs. Initially, these are small vesicles with a red halo, but they soon become hemorrhagic (Figure 6.4). The grayish vesicopustules noted on the fingers are diagnostic. Detection of gonococci in skin lesions is possible with an extensive search, but the diagnosis is facilitated by fluorescent antibody techniques.

Bacillaceae

Bacillus (aerobic) and *Clostridium* (strictly anaerobic) are the two important genera in this family. Both are soil inhabitants capable of inducing serious wound infections in humans and animals. Both genera produce endospores that are extremely resistant to adverse environmental conditions. Clostridia entering the skin through puncture or other deep wounds may produce a spreading necrotic anaerobic infection leading to gangrene.

Anthrax

Bacillus anthracis is usually considered the pathogenic species, although other species have been implicated in infections in compromised hosts. Microbial studies should be performed in a safety hood by someone immunized against the disease.

Specimen collection

A dry swab is used to soak up the fluid in vesicular lesions. With lesions that have developed eschar, a swab moistened with broth is placed beneath the edge of the lesion and rotated, without removal of the eschar. The swab is then placed in a sterile test tube for further tests.

Direct smear

Swabs taken from the lesions are pressed onto slides. For tissue impressions, the freshly cut surface of the tissue is gently pressed onto a slide at several different locations. The smear is air-dried, gently fixed with heat and Gram-stained. *Bacillus anthracis* in infected tissues does not sporulate until exposed to the air.

Morphology

The rods are gram-positive with squarish ends ($3–8 \times 1–1.3\,\mu m$) found singly or in short chains. Spores may be seen as refractile areas (not stained by Gram's method) within the cell. The spores are ellipsoidal, non-budding and centrally located.

Culture

The specimens are inoculated onto blood agar plates and incubated aerobically for 18–24 h at 35–37°C. Colonies are non-hemolytic, 4–5 mm in diameter and off-white to gray in color, with curled edges, giving rise to a Medusa's head appearance. The selective medium is polymyxin–lysozyme EDTA–thallous acetate (PLET) agar.

It is now possible to search routinely for anthrax toxin, and PX01 and PX02 plasmids which code, respectively, for the toxin and capsule. These tests are carried out at special laboratories. Specific polysaccharide epitopes have been identified in *B. anthracis*, and monoclonal antibodies are used in the routine identification of this organism by the US Army Medical Research Institute of Infectious Diseases, Fort Detrick, MD.

Cutaneous anthrax

The lesion begins as a small pruritic pimple and enlarges into a vesicle or ring of vesicles. The typical 'malignant pustule' initially resembles a carbuncle, but soon undergoes necrotic ulceration with a dark and adherent eschar surrounded by satellite pustules. There is usually an extensive area of edema, which may be sufficient to damage the surrounding connective tissues.

Syphilis

This disease is a pathological complex induced by *Treponema pallidum*, a spirochete, and is acquired by sexual contact or by prenatal transmission from the infected mother.

Specimen collection

Prior to obtaining a sample for confirmation of the causal organism, cleansing of the lesion surface is important. This is achieved by compressing the site with a sterile gauze pad dipped in warm physiological saline. For intact lesions, the surface should be carefully scraped, using a sharp scalpel blade. After abrasion, the skin should be kept moist and, if necessary, slight pressure should be applied at the edge of the lesion until some of the fluid is exuded. The fluid is then drawn into a 1-ml syringe containing 0.1–0.2 ml of sterile serum saline (10% normal human or rabbit serum in physiological saline). *Treponema pallidum* is not readily observed in tissue sections from the involved areas during the tertiary stage.

Direct examination

To prepare a wet mount of the sample, the fluid is placed on a slide and covered with a coverslip. This is examined at a magnification of ×400–500 under darkfield illumination. Microscopic observation is not usually achieved by ordinary bacterial or tissue stains because the treponeme diameter is below the resolution of most light microscopes. Darkfield illumination, however, provides an excellent contrast between the microbe and its background.

Treponemes are fine, filamentous, helical cells that are usually found singly. When examined in a wet mount, these microorganisms exhibit vigorous motility.

Serological procedures

Serological tests for syphilis are based on the detection of several different antibodies and are classified by the type of antigen used, such as non-treponemal (reagin) or treponemal. Non-treponemal tests use purified cardiolipin combined with lecithin to detect reagin in the sera of syphilitic patients and, occasionally, in patients with other acute and chronic conditions. Treponemal tests use living or dead *T. pallidum*, or fractions of these treponemes, as antigen sources to detect treponemal antibodies.

Quantitative non-treponemal tests are of value as they establish a baseline of reactivity against which future specimens may be compared. They serve as a guide to distinguish early-latent from late-latent syphilis, and help to differentiate between congenital syphilis and passive reaginemia.

Flocculation is a non-treponemal test for syphilis. Although not absolutely specific for syphilis, it can screen large populations and identify asymptomatic patients. The most widely used flocculation tests are the Venereal Disease Research Laboratory (VDRL) test, the rapid plasma reagin circle card test (RPR-CT) and the automated reagin test (ART). Non-treponemal tests are almost invariably reactive in untreated patients after 4 weeks from the appearance of the chancre, and approximately 25% of infected patients are seroreactive 7 days from the appearance of the chancre. Treponemal tests detect antibodies specific for treponemal antigens, including the fluorescent treponemal antibody absorption (FTA-ABS) test, and the *T. pallidum* immobilization and hemagglutination tests.

The most recent recommendation by the World Health Organization is to screen sera by the VDRL, RPR-CT or automated reagin tests, and to confirm positive sera using the FTA-ABS test.

Tularemia

Francisella (Pasteurella) tularensis is a pleomorphic, gram-negative, aerobic, coccoid / rod-shaped bacterium which requires an enriched medium containing cystine or cysteine.

Wild rodent populations constitute a reservoir of the infection which is transmitted to humans by the bite of ticks or other arthropods, or by direct contact with infected rodents. *Francisella tularensis* is highly infectious for humans and, therefore, extreme precautions should be observed when handling this microorganism.

Specimen collection

The skin around the lesion should be cleansed with alcohol and allowed to dry. The papule is then opened, using a sterile scalpel, and the fluid collected with swabs moistened with broth or saline. Organisms may also be collected from lymph nodes.

Direct examination

Direct examination of the stained smear is of little value due to the small number of microorganisms in the specimen and the difficulty in staining the smear.

Culture

Special media containing cystine or cysteine are required. The addition of penicillin, polymyxin B and nystatin suppresses growth of most normal flora. A large inoculum for primary isolation is highly recommended. After incubation for 36–48 h at 35–37°C, the plates are examined. Colonies appear clear and dewdrop-like on blood-free media, and grayish on blood media. Biochemical characterization is not necessary for routine identification. Usually, rapid slide agglutination with a specific antiserum is adequate. Because there is only one known serological type of *F. tularensis*, a skin test with *F. tularensis* antigen is highly specific and remains positive.

Infection

The clinical manifestations of infection depend on the portal of bacterial entry, such as the skin, eye, or respiratory or gastrointestinal tract. An ulcerated nodule at the point of inoculation is associated with enlargement and tenderness of regional lymph nodes and, later, breakdown of regional lymph nodes. Systemic symptoms and toxemia appear. During the toxemic stage, a generalized maculopapular skin eruption may develop. The clinical picture may resemble erythema multiforme.

Bibliography

Buntin DM, Rosen T, Lesher JL Jr, *et al.* Sexually transmitted diseases: Bacterial infections. *J Am Acad Dermatol* 1991;25:287–99

Fiumara N. The treponematoses. In: Moschella SL, Hurley HJ, eds. *Dermatology.* Philadelphia: WB Saunders, 1992:953–83

Webster SB. Nontreponemal sexually transmitted disease. In: Moschella SL, Hurley HJ, eds. *Dermatology.* Philadelphia: WB Saunders, 1992:987–1007

Chapter 7 Mycobacteria

Mycobacteria are acid-fast non-spore-forming bacteria ('acid-fast' because, once stained, they resist decolorization by acid or alcohol). Thus, they are also not classifiable by Gram-staining because ethyl alcohol containing 3% hydrochloric acid (HCl; acid–alcohol) quickly decolorizes all bacteria except mycobacteria and some *Nocardia*.

Depending on the species and strain, mycobacterial colonies in culture appear from within 2–3 days to up to 2–6 weeks. Those growing rapidly may be cultured on simple media at temperatures varying from 20–40°C; however, pathogenic mycobacteria typically require complex media for growth and have a narrow temperature range. The lipid content of mycobacteria is high compared with other bacteria.

Classification

The modified Runyon classification for atypical mycobacteria is widely used (Table 7.1). Except

Table 7.1 Modified Runyon classification of pathogenic mycobacteria

Cluster	*Species*	*Group*	*Pigment formation*
Mycobacterium tuberculosis complex	Slow-growing pathogens		
	M. tuberculosis–bovis group		
	bacille Calmette–Guérin (BCG)		
	M. africanum		
Non-tuberculous mycobacteria	Slow-growing potential pathogens		
	M. kansasii	I	photochromogen
	M. marinum	I	photochromogen
	M. scrofulaceum	II	scotochromogen
	M. szulgai	II	scotochromogen
	M. avium–intracellulare	III	non-chromogen
	M. haemophilum	III	non-chromogen
	M. ulcerans	III	non-chromogen
	M. xenopi	III	non-chromogen
Mycobacterium fortuitum complex	Rapidly growing potential pathogens		
	M. fortuitum	IV	
	M. chelonae	IV	

Mycobacterium leprae is not included because it has not been grown in culture

for *Mycobacterium leprae*, the mycobacteria are classified into two broad categories: *Mycobacterium tuberculosis*, *M. bovis* and *M. africanum* (*M. tuberculosis* complex); and non-tubercle-inducing bacteria, described on the basis of growth rate and pigment production with and without exposure to light.

Tuberculosis

Morphology

In tissue, tubercle bacilli are slender rods measuring approximately 0.8–5.0 μm. They appear beaded or granular following initiation of therapy. In contrast, coccoid and filamentous forms predominate on artificial media. The Ziehl–Neelsen and Kinyoun acid-fast staining methods are used for identification of acid-fast bacteria. Typical acid-fast bacteria stain red against a blue-stained background (Figure 7.1).

Sample collection and culture processing

Biopsy materials are first minced with sterile scissors and ground in a tissue grinder to a paste-like consistency with sterile saline or 0.2% bovine albumin V. This paste-like material may be further diluted with water. Approximately 0.2 ml of the fluid is removed and inoculated onto 7H10 agar and Löwenstein–Jensen medium.

A direct smear is prepared for acid-fast staining. When contamination due to normal flora is suspected, decontamination is required. The best yield of tubercle bacilli results from the use of the mildest digestion to produce sufficient control of contaminants. Sodium hydroxide, the common digestant, works poorly as a decontaminant because it also inhibits mycobacteria. Trisodium phosphate liquefies sputum rapidly, but requires long exposure for decontamination of the specimen when used alone. Benzalkonium chloride (Zephiran®), used with trisodium phosphate, shortens the time-period required and destroys contaminants. Acetylcysteine-alkali procedures for sputum are also applied toward digestion, and preparation of skin biopsies and bronchial secretions. After treatment, the mixture may then be used to prepare smears for acid-fast staining and inoculation on appropriate media.

Media

Löwenstein–Jensen (egg) medium and 7H10 (clear) medium are standard media. The addition of 0.2% sodium pyruvate improves the growth of *M. bovis*. The addition of L-asparagine (≥0.25%) to 7H10 allows maximum production of niacin.

Except for the rapid growers in Runyon group IV, mycobacteria are slow-growing and require more than one week for visible growth. Cultures need to be examined weekly for 6–8 weeks before being discarded. The more rapid broth systems (such as Bactec™) require only 5–12 days, and rely upon the detection of C^{14}-labeled carbon dioxide produced by growing mycobacteria.

Use of DNA probes with amplification by the polymerase chain reaction (PCR) method has added to the rapid and specific diagnosis of *M. tuberculosis*, allowing detection within hours after sufficient growth has been obtained.

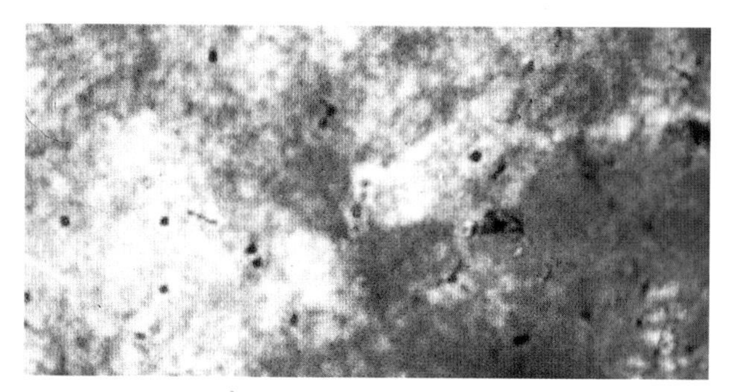

Figure 7.1 Mycobacteria appear as bright red bacilli against a light blue background (acid-fast stain)

Figure 7.2 *Mycobacterium tuberculosis* appears as tan- or buff-colored colonies on Löwenstein–Jensen medium

Culture

Mycobacterium tuberculosis: Colonies of tubercle bacilli are seen on egg media after 2–3 weeks at 37° C. Growth does not occur at either 25° or 45° C. Colonies are small (1–3 mm) in diameter, and dry, rough, granular and buff-colored. Typical colonies have a flat irregular margin and a cauliflower-like center (Figure 7.2). The colonies are difficult to emulsify, and are easily detached from the surface of the medium. These mycobacteria are termed 'eugonic' because of their luxuriant growth.

Mycobacterium bovis: Bovine tubercle bacilli grow more slowly than does *M. tuberculosis*, requiring a longer incubation period (3–6 weeks) at 37° C. The colonies are small at < 1 mm in diameter. They are pale and smooth, and easily emulsified, and adhere to the surface of the medium. As their growth is not luxuriant, they are called 'dysgonic'.

Other mycobacteria ('atypical bacteria')

Runyon's classification of mycobacteria other than tubercle bacilli is widely used. Microorganisms are grouped largely on the basis of chromogenicity and rate of growth. These bacteria are relatively resistant to antituberculosis drugs, and are not transmitted from person to person.

Group I

This group consists of photochromogens, which develop yellow-orange pigment only when the growing culture is exposed to light (Figure 7.3 a); *Mycobacterium marinum* (*M. balnei*), *M. ulcerans* and *M. kansasii* are representative. *Mycobacterium ulcerans* is difficult to cultivate *in vitro* and requires several weeks of incubation at 32° C.

Group II

Comprising the scotochromogens, these present as yellow-orange colonies in either light or darkness (Figure 7.3 b). They are infrequent pathogens in humans. *Mycobacterium szulgai* and *M. scrofulaceum* belong to this heterogeneous group.

Group III

This group comprises non-photochromogens and do not develop pigment when the growing cultures are exposed to light (Figure 7.3 c), for example, the *M. avium–intracellulare* complex. They are characterized by slow growth (10–21 days) at either 37° or 25° C, and the formation of homogeneous, thin, translucent, smooth colonies.

Group IV

This group comprises organisms that produce non-pigmented colonies (Figure 7.3 d) which grow rapidly in simple laboratory media, for example, *M. fortuitum* and *M. chelonae*. These microorganisms produce a visible colony within 3–4 days at ≤37° C. *Mycobacterium fortuitum* is associated with human pulmonary infections and grows on MacConkey

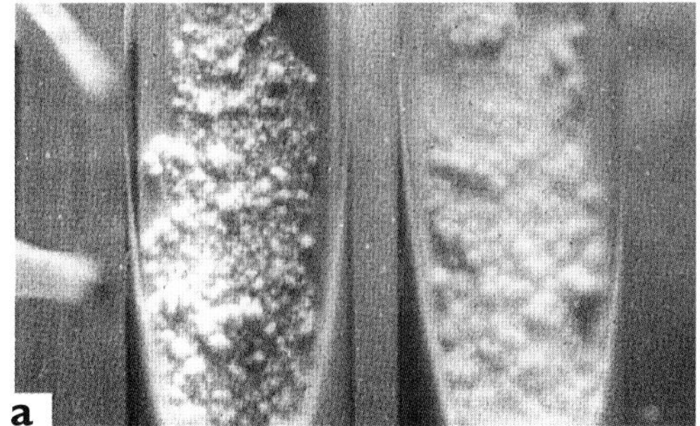

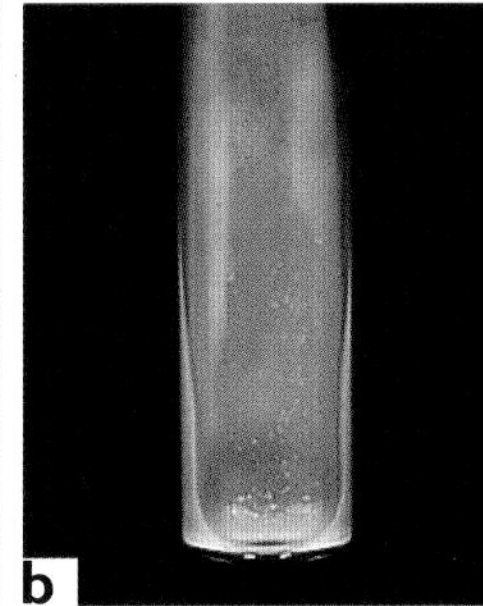

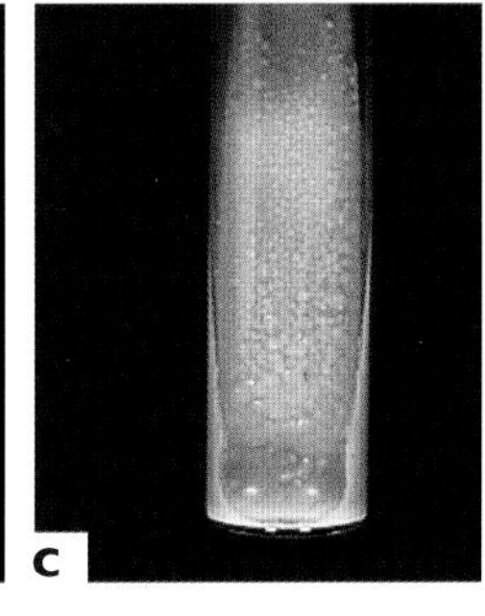

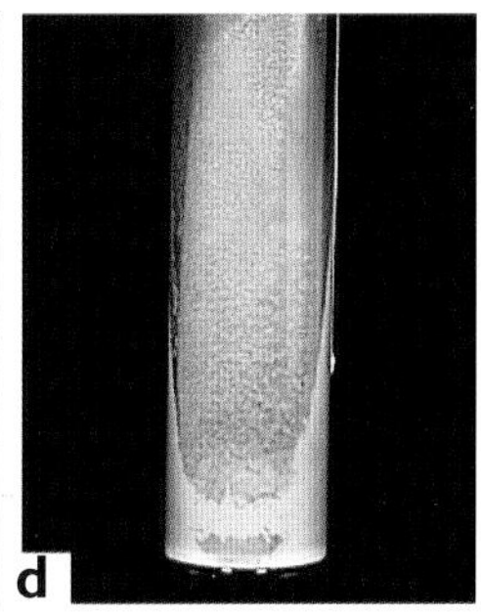

Figure 7.3 Growth of mycobacteria on Löwenstein–Jensen medium in culture tubes incubated at 37° C for 2 weeks: (a) A photochromogen incubated in the dark (left) *vs* in the presence of light (right); (b) a scotochromagen develops yellow pigment regardless of the presence or absence of light; (c) a non-photochromogen produces tan-colored colonies; and (d) a rapid grower also produces tan-colored colonies

agar within 5 days. The accompanying change in the indicator easily differentiates this species from the others, which are inhibited.

Cutaneous infection

Lesions induced by Mycobacterium tuberculosis

Tuberculosis of the skin is produced by accidental inoculation or contamination of a wound in subjects not previously exposed to tuberculosis, or by lymphatic or hematogenous dissemination with focal survival of *M. tuberculosis* in patients who have tuberculosis. Skin tuberculosis in Western countries is rare.

Localized infection

The course of infection begins as an inflammatory nodule (tuberculous chancre, containing bacilli) accompanied by regional lymphangitis and lymphadenitis. Progression of the disease depends on the patient's resistance and effectiveness of treatment.

In immune or partially immune hosts, two major groups of skin lesions may be distinguished: tuberculosis verrucosa; and lupus vulgaris. In the former, inoculation of bacilli incites localized torpid granulomatous papules or verrucous nodules containing few bacilli. Lupus vulgaris begins early in life and often on the face. The patchy lesions are studded with soft confluent tubercles which, on diascopy, appear as small, yellowish-brown, 'apple-jelly' nodules. New nodules develop at the periphery whereas the center undergoes atrophic scarring. Ulceration, edema and hypertrophy due to lymphatic obstruction develop. Tubercle bacilli are scarce. In temperate climates, the majority of lupus lesions are on the face whereas those of tuberculosis verrucosa are on the hands. In tropical areas, the distribution of lesions may be different.

In scrofuloderma, tuberculosis of the lymph nodes or the bones extends into the skin, resulting in the development of ulcers. Numerous fistulas may communicate beneath ridges of bluish-colored skin. Tubercle bacilli are usually isolated from the pus.

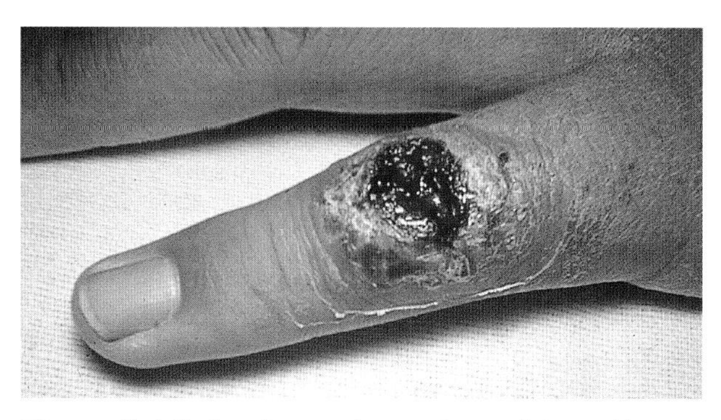

Figure 7.4 Swimming-pool granuloma due to *M. marinum*. Courtesy of Dr Axel Hoke, UCSF, San Francisco, CA

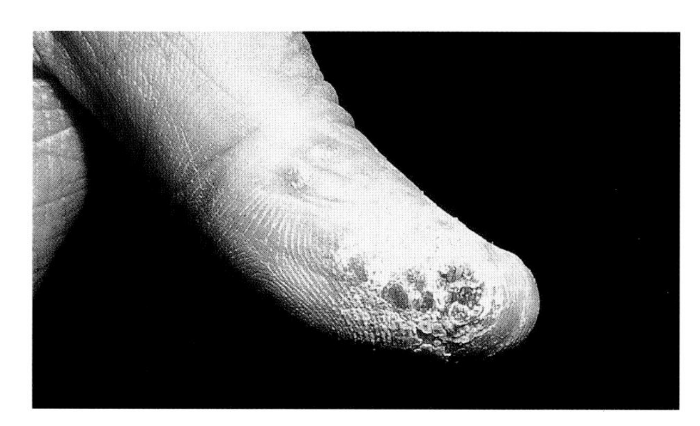

Figure 7.5 Infection of the thumb due to *M. marinum*. Courtesy of Dr Axel Hoke, UCSF, San Francisco, CA

Disseminated infection

Bacteria may spread by the hematogenous route in patients with fulminating tuberculosis, and passive defense takes the form of miliary tuberculosis in the skin. These lesions are teeming with tubercle bacilli. When hypersensitivity to tubercle bacilli is present, their antigens, distributed *via* the bloodstream, cause the appearance of tuberculids, as seen in lichen scrofulosus. The strong immunological reaction of tuberculids destroys mycobacteria, resulting in strongly positive tuberculin reactivity.

Lesions induced by mycobacteria other than Mycobacterium tuberculosis

Mycobacterium marinum

Many cases of infection due to *M. marinum* are seen in children and adolescents who frequent swimming pools or clean fish tanks. There is often a history of trauma but, even in its absence, the lesions are typically at sites most often exposed to injury [fingers (Figures 7.4 and 7.5), knees, elbows or bridge of the nose]. The usually solitary lesions are tuberculoid granulomas that rarely show acid-fast microorganisms; nevertheless, the skin tuberculin test is positive. Hot compresses may be curative. Disseminated infection to other skin areas may occur in immunocompromised patients.

Mycobacterium ulcerans

In some tropical regions, chronic ulcers caused by this bacterium are common. Lesions are usually on the arms or legs, but may occur elsewhere (except palms and soles). Most patients have a single painless cutaneous ulcer with characteristic undermined edges. Geographical association of the disease with swamps and water courses has been reported. On primary isolation, *M. ulcerans* may not grow at 37° C. The preference for a lower temperature may explain the restriction of lesions to the skin and subcutaneous tissues. *Mycobacterium ulcerans* is sensitive to a number of drugs *in vitro*, but the clinical results of chemotherapy have been disappointing.

Mycobacterium scrofulaceum

This microorganism is widely distributed. It is seen in children who are accidentally infected while playing. The portal of entry is typically the oropharynx, with development of cervical lymphadenitis that later extends to form a draining sinus tract (Figure 7.6).

Mycobacterium fortuitum–Mycobacterium chelonae *complex*

These are common pathogens among immunosuppressed patients, and may be differentiated from other mycobacteria on the basis of their rapid growth in culture. They are widely distributed and commonly found in water. Skin lesions may be the result of direct inoculation, extension from underlying tissues or dissemination (for example, vascular spread from the lungs, lymph nodes or endocardium). Drained, and especially expressed, material is positive for acid-fast bacilli, making diagnosis rapid.

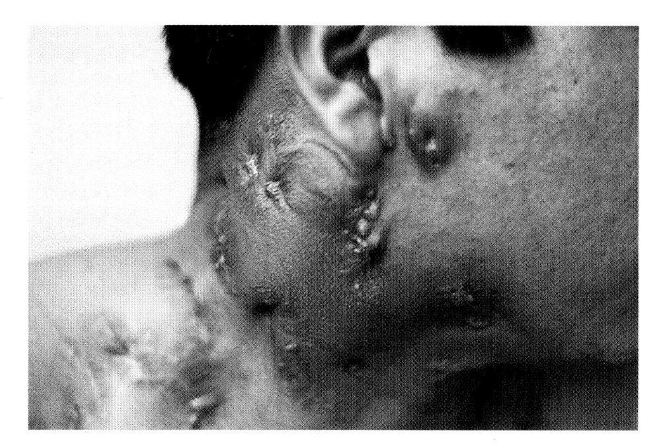

Figure 7.6 Scrofuloderma due to *M. scrofulaceum* involving the cervical area. Courtesy of Dr Axel Hoke, UCSF, San Francisco, CA

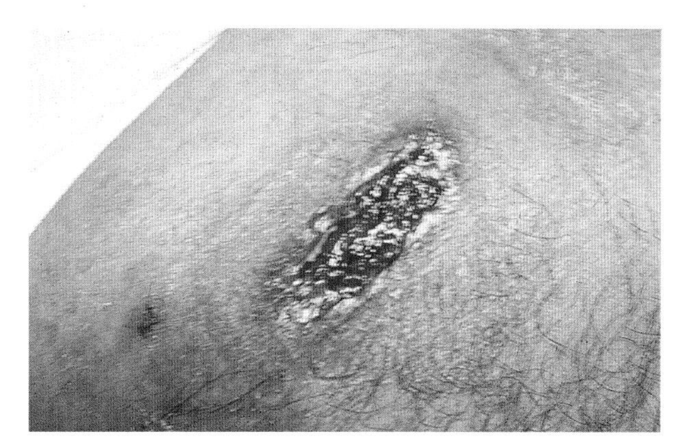

Figure 7.7 Skin lesion due to *M. haemophilum* infection in a patient with AIDS. Courtesy of Dr Timothy Berger, San Francisco, CA

Mycobacterium avium–intracellulare *complex (MAI or MAC)*

The optimal temperature for the growth of these bacteria is 41° C, and colonies are smooth and nonpigmented. These mycobacteria have been found in water, soil, food and animals. Infection due to MAI is uncommon in immunocompetent subjects. However, in the USA, disseminated MAI is a common opportunistic infection. Another mycobacterium frequently associated with skin lesions in immunocompromised patients is *Mycobacterium haemophilum* (Figure 7.7). Diagnosis is made on the basis of culture from blood or tissue.

Mycobacterium leprae

This acid-fast bacterium is present in lepromatous tissue, and cannot be cultivated on either artificial media or human tissue culture cells. However, the microorganism has been grown in the footpads of irradiated thymectomized mice. Humans and armadilloes are the only hosts; there is no soil reservoir.

Direct smear

A sample is taken by scraping the skin (dermis, not epidermis) and mucous membrane of, for example, the nasal septum in patients with the lepromatous form of the disease. When stained by the Ziehl–Neelsen method, large numbers of acid-fast bacilli packed within lepra cells lying in parallel bundles may be seen. The bacilli are also found within the endothelial cells of blood vessels.

Infection

Primarily affecting the skin, mucous membranes, peripheral nerves and the eyes, leprosy occurs in three forms, probably depending on the immunological status of the patient:

(1) The lepromatous form is characterized by nodular lesions containing numerous bacilli. Nodules of the earlobes, extremities and trunk may ulcerate and develop secondary infection;

(2) The tuberculoid form is associated with sparse macular lesions that are insensitive to pain, heat and touch. Few or no bacilli are seen in the lesions;

(3) The intermediate (indeterminate) form is characterized by macular lesions that demonstrate bacilli.

Antibiotic-sensitivity testing

Although it is advisable to test the sensitivity of mycobacteria to antibiotic compounds, it is usually not necessary to wait for the results of these tests before initiating therapy. If a patient does not respond to therapy, it is possible that drug-resistant bacilli have emerged, and testing for susceptibility to antibiotics should be repeated.

Bibliography

Beyt BE Jr, Ortbals DW, Santa Cruz DJ, *et al.* Cutaneous mycobacteriosis: Analysis of 34 cases with a new classification of the disease. *Medicine (Baltimore)* 1981;60:95–109

Kakahel KU, Fritsch P. Cutaneous tuberculosis. *Int J Dermatol* 1989;28:355–62

Roberts GD, Elmer W, Koneman EW, Kim YK. Mycobacterium. In: Ballows A, ed. *Manual of Clinical Microbiology*. Washington, DC: American Society of Microbiology, 1991:304–39

Sehgal V, Wagh SA. Cutaneous tuberculosis: Current concepts. *J Dermatol* 1990;29:237–52

Section II Ectoparasites, arthropods and pinworms
by Dirk M. Elston

Chapter 8	**Arthropods and human disease**	**46**
Chapter 9	**Arachnida**	**48**
Chapter 10	**Chilopoda and Diplopoda**	**56**
Chapter 11	**Insecta**	**57**
Chapter 12	**Helminthic pathogens (worms)**	**65**

The views here expressed are those of the author, and are not to be construed as official, or as representing those of the Army Medical Department, the US Air Force or the Department of Defense.

Chapter 8 Arthropods and human disease

The phylum Arthropoda includes insects, arachnids, centipedes, millipedes and crustaceans. All classes of arthropods are implicated in human disease. Arthropods serve as vectors of viral, bacterial, rickettsial and helminthic pathogens. Even crustaceans may cause skin disease in the form of contact urticaria and sea-bather's eruption[1].

Individuals vary in their attractiveness to insects that bite. Any given member of a family may manifest a reaction to biting arthropods that is far out of propor-tion compared with those of other family members. To a large extent, this is a measure of their immune response to the bite. Some patients who insist that insects 'like them better' than others in the family are supported in this claim by scientific evidence. For some insects, much of a subject's attractiveness depends upon breath carbon dioxide levels[2].

Identification of arthropods: Methods and general principles

Standard operating procedure

Definition Gross and direct light-microscopic examination to identify ectoparasites.

Principle of the test Some ectoparasites or biting / stinging arthropods are brought to the dermatologist for identification. Others are found on the skin or hair by means of simple techniques. Visual identification can establish the diagnosis of scabies or pediculosis. Arthropod identification may be important in determining the appropriate treatment for envenomation, and in the control of vector-borne diseases.

Clinical indications Identification of an arthropod which has bitten or stung a subject; evaluation of scalp pruritus, perianal pruritus, generalized pruritus or pruritus localized to the web spaces, axillae, umbilicus and / or groin; and surveillance of disease vectors

Equipment Gloves, forceps, glass slides and coverslips, scalpel blades, India ink, saline, mineral oil, 20% potassium hydroxide (KOH) solution and a light microscope.

Procedure Larger arthropods are examined grossly. 'Hard-shelled' arthropods are preserved in absolute alcohol, 70% ethanol or 70% isopropyl alcohol until they are examined. Alcohol fixation renders the arthropod harmless, as they may be dangerous and should not be handled or examined while alive. 'Soft' larval forms of arthropods, such as maggots, may need special handling. As they tend to contract in alcohol, contraction may be minimized by killing the maggot in hot water before alcohol fixation.

Small arthropods, such as lice, mites and small flies, may be immobilized in a drop of immersion or mineral oil for examination under low-power microscopy. Ether or chloroform may be used to kill flying insects prior to mounting. After placing a drop of oil on a slide, the ectoparasite is placed in the oil and covered with a coverslip. Normal saline, histological mounting media or 20% KOH solution may also be used, and are more suitable for examination of ova.

Interpretation Refer to the photographs and descriptions presented in Chapters 9–12. Many arthropods are easily identified but, if the clinician has any uncertainty, the specimen should be forwarded to a large clinical laboratory, medical entomologist or state laboratory. The state or local health department may be able to refer the clinician to a knowledgeable individual at a local university or government agency.

References

1. Freudenthal AR, Joseph PR. Sea bather's eruption. *N Engl J Med* 1993;329:542–4

2. Schofield SW, Sutcliffe JF. Human individuals vary in attractiveness for host-seeking black flies (Diptera: Simuliidae) based on exhaled carbon dioxide. *J Med Entomol* 1996;33:102–8

Additional sources of information

Textbooks

Fritsche TR, Pfaller MA. Arthropods of medical importance. In: Murray PR, Baron EJ, Pfaller MA, *et al.*, eds. *Manual of Clinical Microbiology.* Washington, DC: American Society of Microbiology, 1995:1257–74

Noble ER, Noble GA, Schad GA, MacInnes A. *Parasitology: The Biology of Animal Parasites.* Philadelphia: Lea & Febiger, 1989

Russell FE. Toxic effects of animal toxins. In: Klaassen CD, ed. *Toxicology.* New York: McGraw–Hill, 1996: 841–54

Periodicals

Allen C. Arachnid envenomations. *Emerg Med Clin N Am* 1992;10:269–98

Cavagnol RM. The pharmacological effects of hymenoptera venoms. *Ann Rev Pharmacol Toxicol* 1977;17: 479–98

Groff JW. Organisms and associated diseases. *J Assoc Military Dermatol* 1983;9:72–5

Kurgansky D, Burnett JW. Diptera mosquitoes. *Cutis* 1988;41:317–8

Millikan LE. Mite infestations other than scabies. *Semin Dermatol* 1993;12:46–52

Modley CE, Burnett JW. Tick-borne dermatologic diseases. *Cutis* 1988;41:244–5

Parish LC, Schwartzman RM. Zoonoses of dermatological interest. *Semin Dermatol* 1993;12:57–64

Reisman RE. Insect stings. *N Engl J Med* 1994;331:523–7

Wilson DC, King LE. Spiders and mites. *Dermatol Clin* 1990;8:277–86

Chapter 9 Arachnida

Mites

Sarcoptes scabiei

The adult female mite burrows through the stratum corneum to the level of the upper stratum granulosum. Lesions characteristically involve the web spaces, wrists, areolae, glans penis and umbilicus. The diagnosis is confirmed by the demonstration of an adult mite, eggs or feces in scrapings from a burrow (Figure 9.1). Burrows may be highlighted by a drop of India ink applied to the skin. Fluorescence microscopy (Figure 9.2) is helpful in identifying mites and ova[1].

Norwegian (crusted) scabies may be seen in disabled, elderly, mentally handicapped or immunosuppressed patients. These patients are infested with innumerable mites and are highly contagious.

Demodex folliculorum

Demodex mites are normal human fauna. *Demodex folliculorum* (Figure 9.3) usually involves the face whereas *D. brevis* commonly infests the chest and back. *Demodex* mites have been implicated as a possible cause of acne rosacea. In immunocompromised hosts, they may overpopulate and cause a dermatitis[2]. *Demodex* mites may also cause demodectic alopecia ('human mange').

Zoonotic mites

A wide variety of zoonotic mites are able to affect

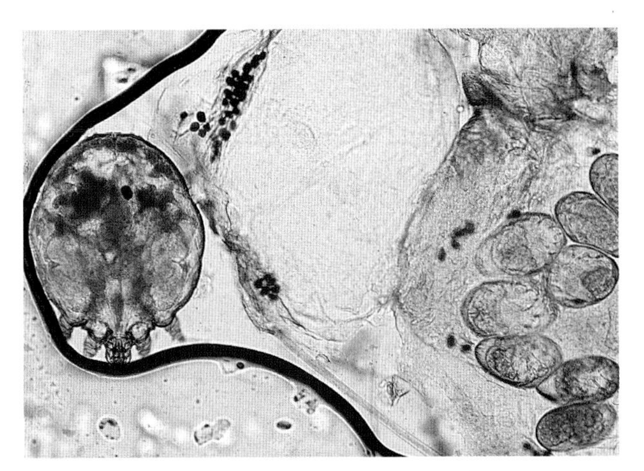

Figure 9.1 *Sarcoptes scabiei.* The diagnosis of scabies is made by identifying the adult mites, ova or feces

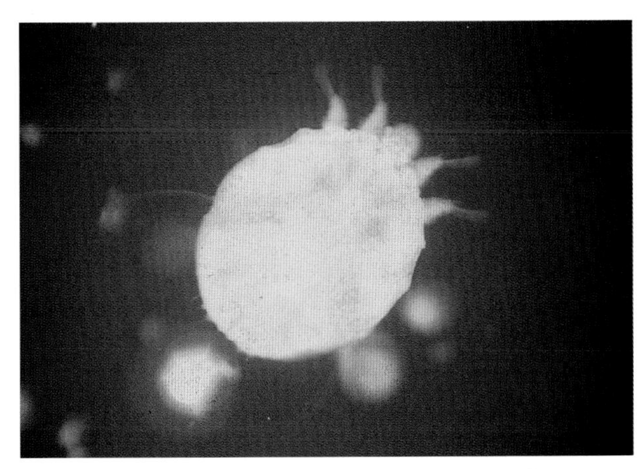

Figure 9.2 Autofluorescence renders the scabies mite, ova and feces easy to find with fluorescence microscopy

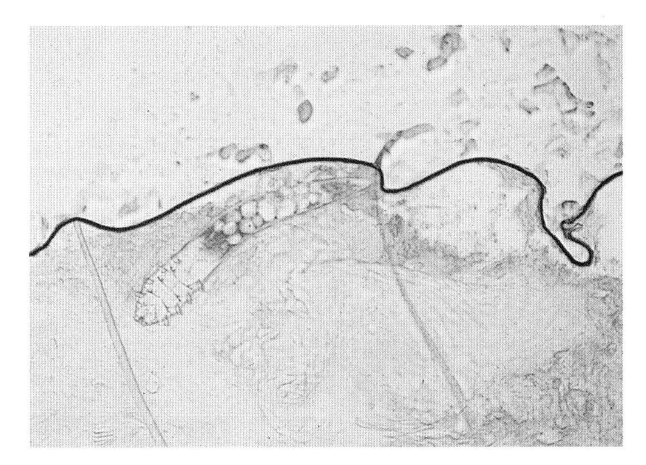

Figure 9.3 *Demodex folliculorum*, one of the normal human fauna

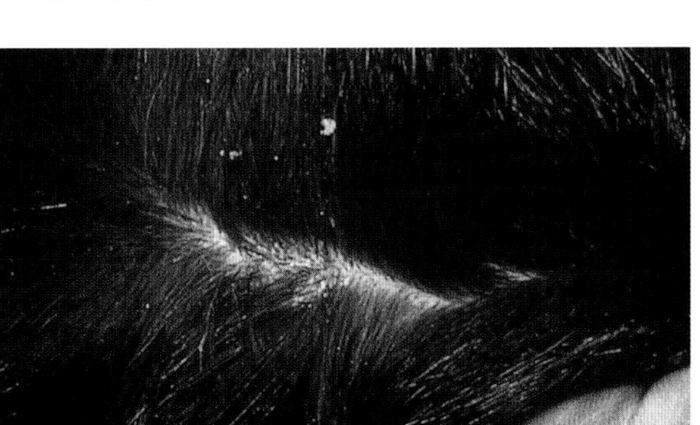

Figure 9.4 *Cheyletiella* dandruff on a cat. Courtesy of Dr Joseph Cvancara, San Antonio, TX

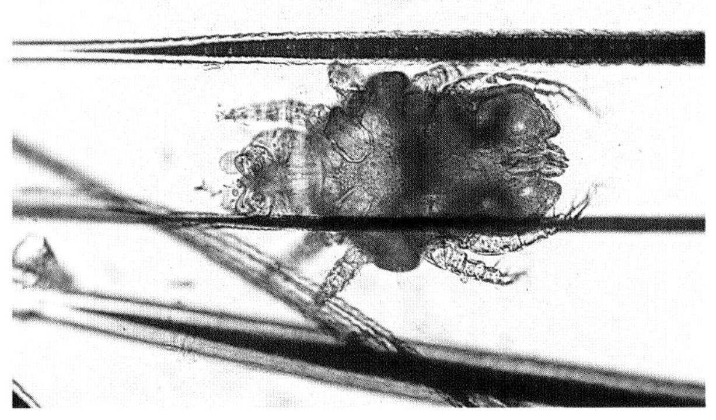

Figure 9.5 *Cheyletiella blakei*. Courtesy of Dr Joseph Cvancara, San Antonio, TX

humans. Establishing the diagnosis of zoonotic mite infestation can spare the patient from needless and, indeed, useless treatments for scabies. Eradication of the environmental source will cure the patient.

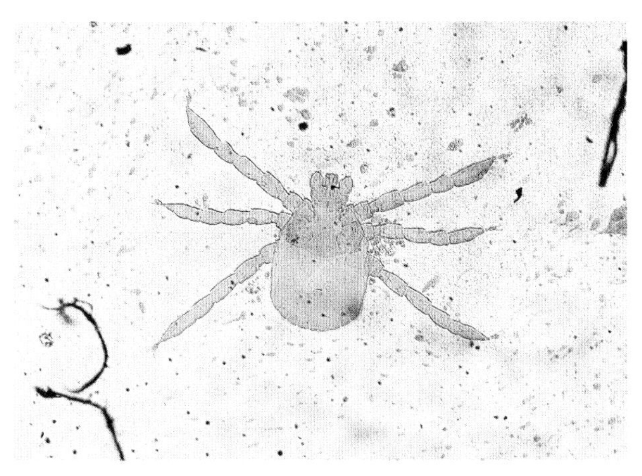

Figure 9.6 Larval mites are six-legged whereas adult mites have eight legs

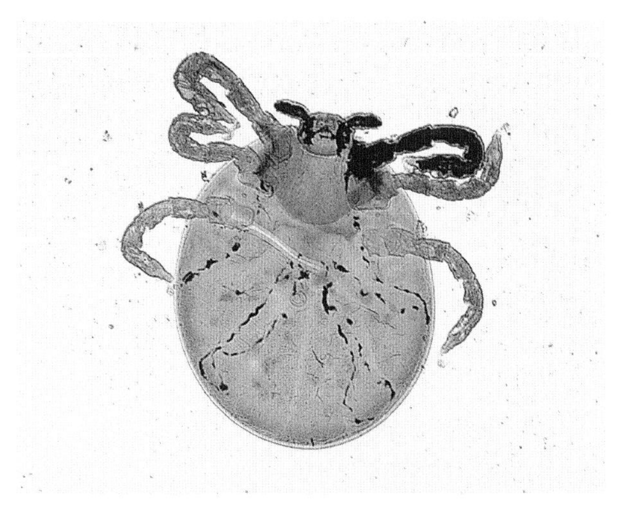

Figure 9.7 Larval chigger mite

Prostigmata mites

Cheyletiella mites have a prominent hooked claw distally on each palp. They may infest mammals and birds, and are a frequent cause of pet-associated dermatosis (Figures 9.4 and 9.5). Commonly, the pet appears to have dandruff[3].

Trombicula alfreddugèsi (Trombidiidae mites, chiggers) cause bite reactions and transmit scrub typhus (*Rickettsia tsutsugamushi*). Bites are characteristically grouped on the lower extremities and are intensely pruritic. The larval stage is identified by the presence of six, rather than eight, legs (Figures 9.6 and 9.7).

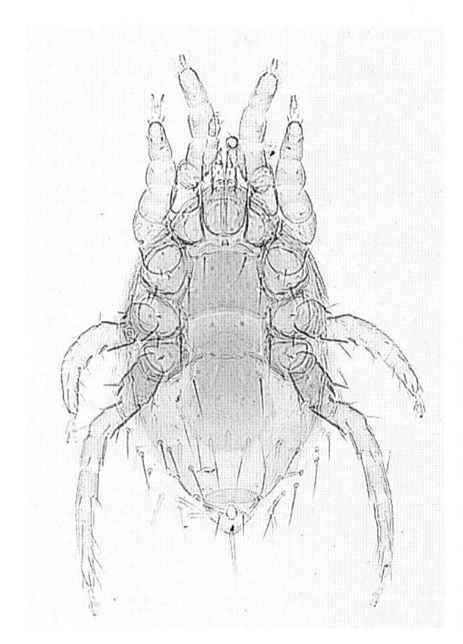

Figure 9.8 *Laelaps echidnina* is a rat mite

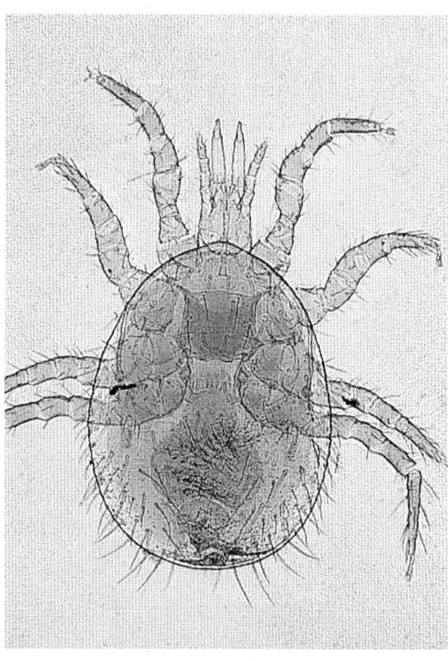

Figure 9.9 *Laelaps castroi*. This genus comprises tick-like mesostigmata mites

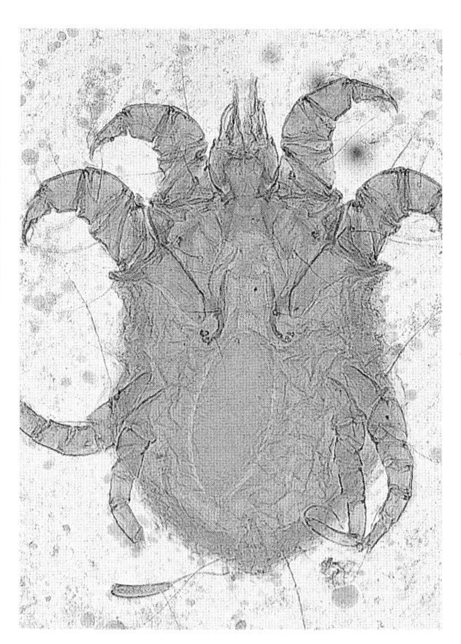

Figure 9.10 *Psoroptes communis*. Larger than *Sarcoptes* mites, this genus causes 'scab' in livestock

Mesostigmata mites

These zoonotic mites are closely related to and resemble small ticks, and include *Ornithonyssus bacoti*, a rat mite, and *Liponyssoides* (formerly *Allodermanyssus*) mouse mites, the vector for rickettsial pox. *Ophionyssus* snake mites, *Dermanyssus* poultry mites and *Echinolaelaps* (*Laelaps*) rodent mites (Figures 9.8 and 9.9) are also included in this suborder.

Astigmata mites

These include animal scabies mites and other mange mites, such as *Otodectes* and *Chorioptes*. Sustained human infestation with animal scabies (mange) mites may occur[4]. *Psoroptes* mites (Figure 9.10) affect livestock and domestic animals. *Dermatophagoides* mites infest birds' nests and include the common house dust mite. *Glycyphagus* mites include the 'grocer's itch' mite.

Ticks

In the USA, ticks are the major vectors of arthropod-borne diseases[5]. Ticks of dermatological importance include hard ticks (Ixodidae; Figure 9.11) and soft ticks (Argasidae). Hard ticks are identified by their hard scutum (dorsal plate) and anteriorly projecting mouth parts. *Dermacentor* and *Amblyomma* ticks have ornate markings on the scutum. In male ticks, the scutum covers the entire body whereas female ticks have a small scutum to allow distention of the body when engorged with blood.

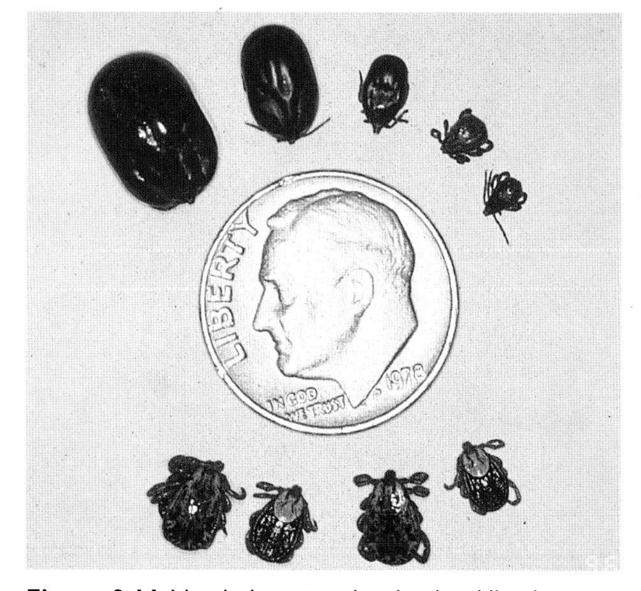

Figure 9.11 North American hard ticks. All ticks progress through larval, nymph and adult stages. Larval forms have six legs

The presence of a small hard scutum and anterior mouth parts helps distinguish an engorged female hard tick from a soft tick. Hard ticks remain attached to their hosts for a long period of time and are therefore more frequently brought to the dermatologist's attention than are soft ticks.

Tick removal

With *Ixodes*, *Dermacentor* and *Amblyomma* ticks, grasping them as closely as possible by their mouthparts and pulling straight upwards, using even pressure, frequently results in complete removal of the tick[6,7]. If the mouthparts break off as a result of pulling, the embedded fragments may be larger than the small fragments consistently left after rotation without pulling[7]. Applications of petrolatum, fingernail polish, alcohol or hot matches are generally ineffective[7].

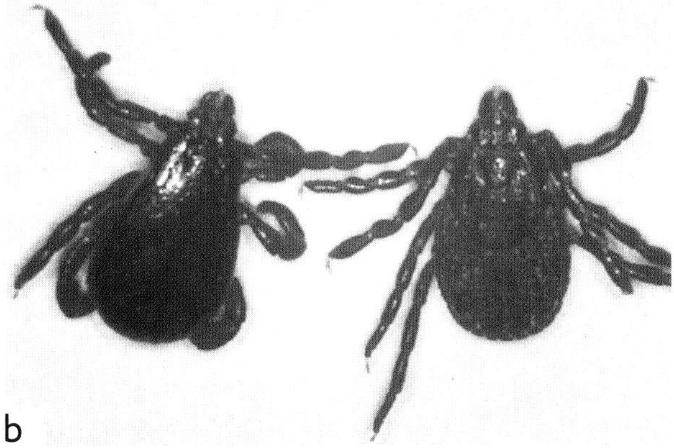

Figure 9.12 *Ixodes scapularis* (formerly *I. dammini*): (a) The engorged females are differentiable from soft ticks by the anterior mouth parts and small, hard, anterior dorsal scutum; (b) the non-engorged teardrop shape is lost when engorged. Black legs and an inornate scutum are characteristic of *I. scapularis*

Ixodes ticks (black-legged ticks)

Ixodes ticks have a characteristic teardrop shape and an inornate scutum, and an anal groove visible anterior to the anus. In other hard ticks, the anal groove is posterior to the anus and may be inconspicuous. Usually, female ticks are found attached to the host (Figure 9.12).

Ixodes ticks are vectors of Lyme disease and babesiosis. Recent evidence suggests that the Lyme disease spirochetes may be transmitted between mating pairs of *Ixodes* ticks either as a sexually transmitted disease or by vampirism (sucking blood from a mate)[8]. *Ixodes* ticks have been associated with tick paralysis[9]; prompt removal of the ticks usually results in full recovery[9].

Dermacentor andersoni (wood tick)

Common in the Rocky Mountain states of the USA, these ticks have a highly ornate scutum and short mouthparts attached to a rectangular base (Figure 9.13). They carry Rocky Mountain spotted fever, tularemia, ehrlichiosis and Colorado tick fever, and are a cause of tick paralysis.

Figure 9.13 *Dermacentor andersoni* (wood tick). The female tick (left) has a smaller scutum than the male tick (right)

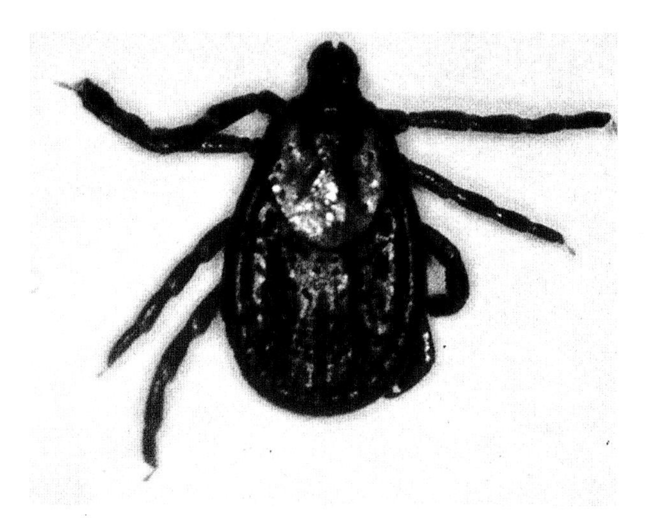

Figure 9.14 *Dermacentor variabilis* (dog tick) female

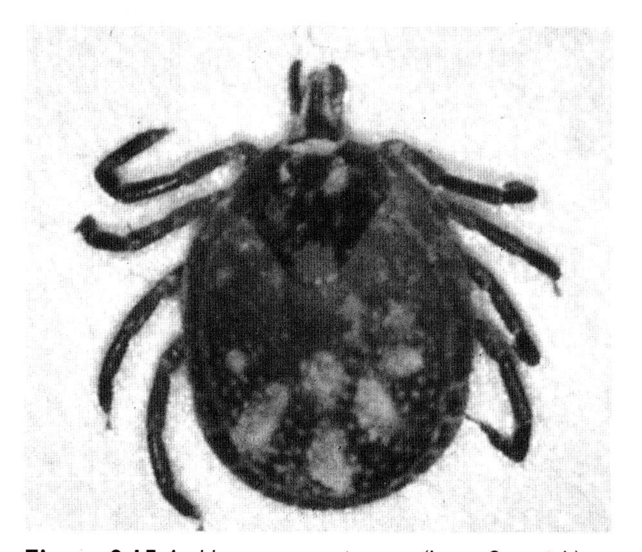

Figure 9.15 *Amblyomma americanum* (Lone Star tick)

Dermacentor variabilis (dog tick)

Common throughout the USA, except in the Rocky Mountains, these ticks are characterized by an ornate scutum and short mouthparts attached to a rectangular base (Figure 9.14). *Dermacentor variabilis* is the primary vector for Rocky Mountain spotted fever, and is also a vector for tularemia and a cause of tick paralysis.

Amblyomma americanum (Lone Star tick)

Recognized by the ornate 'lone star' dot on the scutum, these ticks have long mouthparts and an ornate scutum (Figure 9.15). *Amblyomma* ticks carry

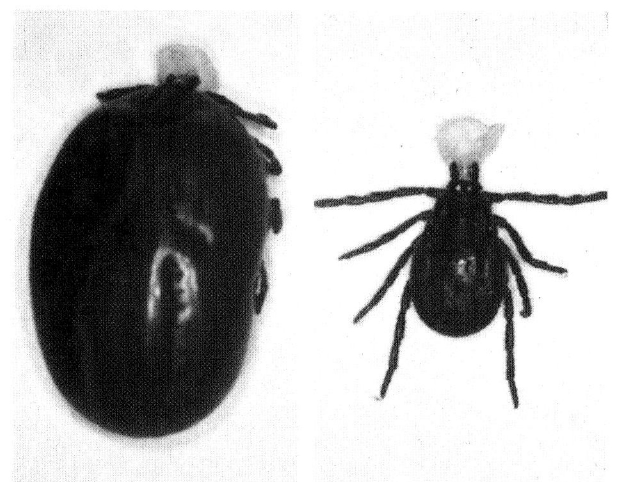

Figure 9.16 *Rhipicephalus sanguineus* (brown dog tick). Engorged (left) and non-engorged (right) appearances

Figure 9.17 *Ornithodoros* ticks are soft ticks

Rocky Mountain spotted fever, tularemia and ehrlichiosis, and are a cause of tick paralysis.

Rhipicephalus sanguineus (brown dog tick)

These have an inornate scutum and short anterior mouthparts attached to a widened, somewhat diamond-shaped, base (Figure 9.16). They transmit Rocky Mountain spotted fever, ehrlichiosis and boutonneuse fever. Other hard ticks, such as *Boophilus*, are pests to cattle and other livestock, but rarely affect humans.

Ornithodoros

A vector of borrelial relapsing fever, these soft ticks (Figure 9.17) usually live in animal nests and take

Figure 9.18 Brown recluse spider. Note the violin-shaped mark on the cephalothorax

Figure 9.19 Black widow spider. Note the shiny black bulbous abdomen

frequent, small, blood meals from their hosts. They do not remain attached to their hosts and are, therefore, infrequently brought to the attention of dermatologists.

Spiders

Brown recluse spiders

Loxosceles reclusa, the brown recluse spider, is identified by the violin-shaped mark on its dorsal cephalothorax (Figure 9.18). This is a hunting spider that is often found in closets, attics and other sites where 'junk' is stored. These spiders are found throughout the Americas, but most bites are reported from Arkansas, Missouri, Kansas and Tennessee.

Black widow spiders

Latrodectus mactans is identified by its bulbous shiny black abdomen with the characteristic red hourglass-shape on the ventral surface (Figures 9.19 and 9.20). They are web spinners, and often found in woodpiles and under outhouse seats. The neurotoxin contained in the venom causes symptoms mimicking an acute abdomen. Treatments include calcium gluconate and muscle relaxants (diazepam). A horse serum antivenom is available.

Figure 9.20 Black widow spider. Note the red hourglass-shape on the ventral surface

Tarantula spiders

Members of the Theraphosidae family (tarantulas) are large hairy spiders (Figure 9.21) commonly

Figure 9.21 Tarantula hairs produce urticarial reactions

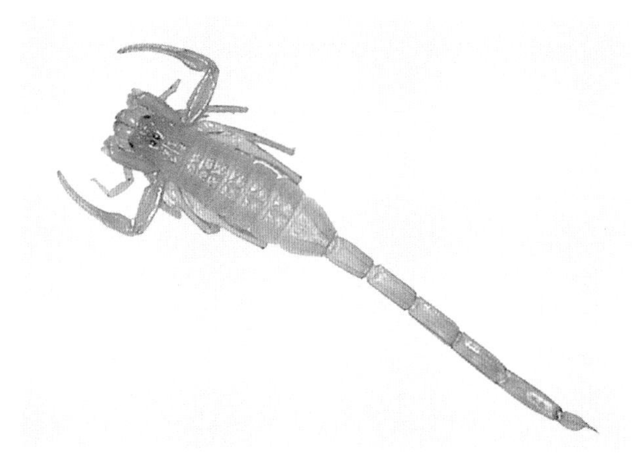

Figure 9.22 *Centruroides exilicauda* (*C. sculpturatus*) is the only deadly scorpion found in the USA

Figure 9.23 Other scorpions found in the USA are seldom life-threatening

Figure 9.24 This scorpion was found in a Florida 'tent city' following Hurricane Andrew, which left many Florida residents without the doors and windows usually separating them from Florida's numerous arthropods

found in the southwestern USA. Contact with the hairs produces an urticarial reaction[10]. Tarantulas are also capable of flicking bristles into the eyes of their enemies[11].

Scorpions

Centruroides exilicauda (*C. sculpturatus;* Figure 9.22), found in Arizona, is the only deadly scorpion in the USA. Antivenom is available, but may result in serum sickness[12]. Other treatment is largely supportive. The stings of other American scorpions (Figure 9.23) are painful but, except in cases of allergic anaphylactic reactions, are not life-threatening[13].

Scorpions typically enter houses which have stood empty for a period of time. When new occupants arrive, the scorpions tend to move out. They also enter houses when their environment has been disturbed (Figure 9.24). Southwesterners frequently find scorpions in their homes when a house is being built in an adjacent lot.

Scorpions commonly settle in shoes during the night. Campers in the southwestern area are well-advised to shake out their shoes in the morning.

References

1. Bhutto AM, Honda M, Kubo Y, *et al.* Introduction of a fluorescence-microscopic technique for the detection of eggs, eggshells, and mites in scabies. *J Med Entomol* 1996;33:102–8

2. Ivy SP, Mackall CL, Gore L, *et al.* Demodicosis in childhood acute lymphoblastic leukemia: An opportunistic infection occurring with immunosuppression. *J Pediatr* 1995;127:751–4

3. Blankenship M. Mite dermatitis other than scabies. *Dermatol Clin* 1990;8:265–75

4. Mitra M, Mahabta SK, Sen S, *et al.* *Sarcoptes scabiei* in animal spreading to man. *Trop Geo Med* 1993;45:142

5. Spach DH, Liles WC, Campbell GL, *et al.* Tick-borne diseases in the United States. *N Engl J Med* 1993; 329:936–47

6. De Boer R, van den Bogaard AEJM. Removal of attached nymphs and adults of Ixodes ricinus (Acari: Ixodae). *J Med Entomol* 1993;30:748–52

7. Needham G. Evaluation of five popular methods for tick removal. *Pediatrics* 1985;75:997–1002

8. Alekseev AN, Dubinina HV. Exchange of *Borrelia burgdorferi* between *Ixodes persulcatus* (Ixodidae: Acarina) sexual partners. *J Med Entomol* 1996;33:351–4

9. Haas E, Anderson D, Neu R, *et al.* Tick paralysis – Washington, 1995. *MMWR* 1996;45:325–6

10. Cooke JAL, Miller FH, Grover RW, *et al.* Urticaria caused by tarantula hairs. *Am J Trop Med Hyg* 1973; 22:130–3

11. Edwards WC. Response to the Letter to the Editor on the spitting spider from Hell. *Vet Hum Tox* 1990; 32:330

12. Carbonaro PA, Janniger CK, Schwartz RA. Scorpion sting reactions. *Cutis* 1996;57:139–41

13. Allen C. Arachnid envenomations. *Emerg Med Clin N Am* 1992;10:269–98

Chapter 10 Chilopoda and Diplopoda

Centipedes

Centipedes have flat long bodies and 15 or more pairs of legs (Figure 10.1). They move quickly and typically hunt at night. All have a poisonous bite, although many are too small to be great danger to humans.

Figure 10.1 Centipede. Note the flat body and long antennae

Millipedes

Millipedes have numerous body segments, each with two pairs of legs (Figure 10.2). Unlike centipedes, their bodies are not flat, and they have short antennae. Millipedes are slow-moving and eat decaying plant material. They lack the poisonous bite of centipedes, but are capable of elaborating an irritating foul-smelling fluid.

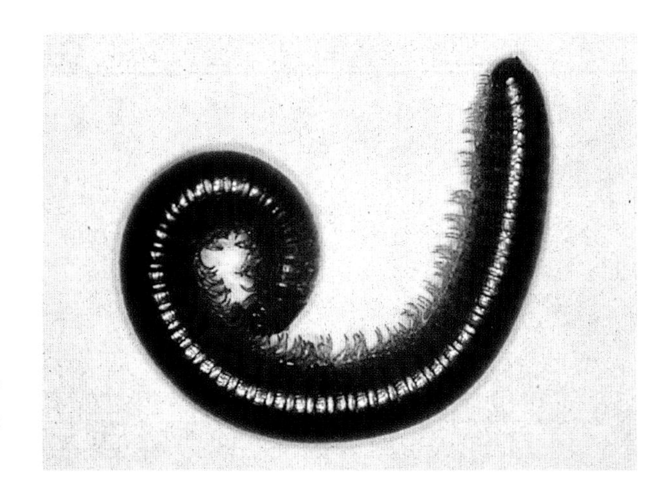

Figure 10.2 Millipede. Note the tubular body, the numerous pairs of legs and short antennae

Chapter 11 Insecta

Phthiraptera (lice): Anoplura and Mallophaga

Sucking lice (Anoplura) and chewing lice (Mallophaga) are wingless parasites which infest humans and animals. Humans may become secondary hosts to 'non-human' lice. Sucking lice have small heads with piercing mouthparts whereas chewing lice have broad heads with well-developed mandibles. Lice attach their egg cases (nits) to hair, feathers or clothing.

Pthirus pubis (pubic or crab louse)

Pubic lice are identifiable by their crab-like appearance (Figure 11.1). Their eggs (nits) are found attached to pubic hairs or to eyelashes. Pubic lice are associated with maculae ceruleae (steel-gray spots) in the pubic region.

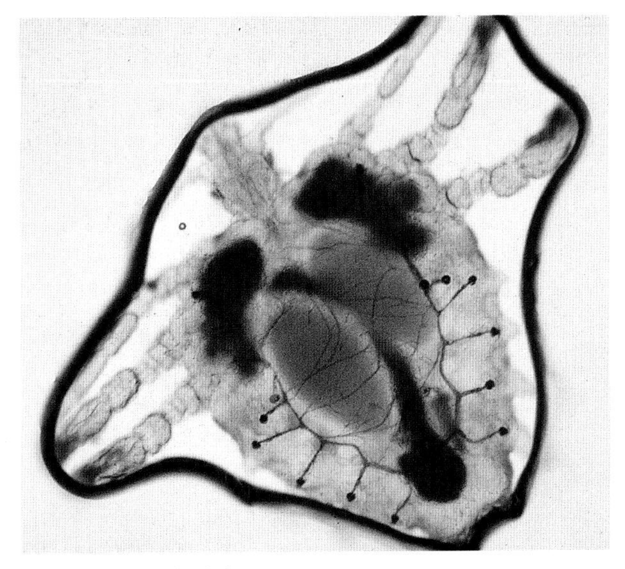

Figure 11.1 Crab louse

Crab lice may also infest the scalp[1]. Identification of crab louse scalp infestation is clinically significant. Effective treatment of crab lice (compared with head lice) may require a higher concentration of permethrin. Many clinicians use the 5% cream rather than the 1% cream rinse.

Eyelash nits are treated with white petrolatum, fluorescein dye or yellow mercuric oxide ointment[2].

Pediculus humanus capitis (head louse)

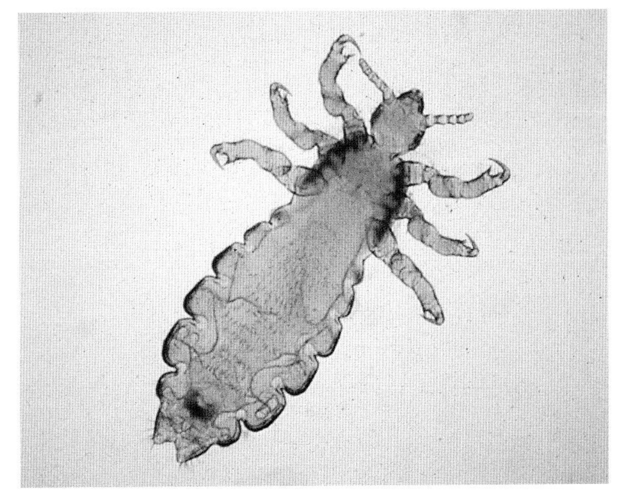

Figure 11.2 Head louse

Head lice have an elongated body (Figure 11.2). The

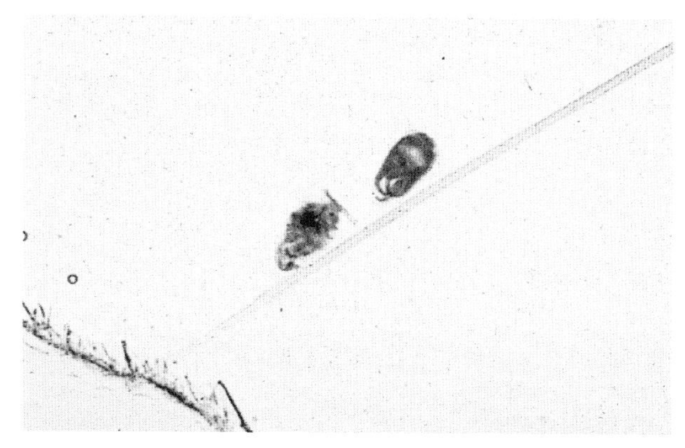

Figure 11.3 Nits are flask-shaped egg cases 'cemented' to hair shafts (crab and head lice) or clothing fibers (body lice)

Figure 11.4 Body louse

Figure 11.5 *Bovicola bovis* (biting cattle louse)

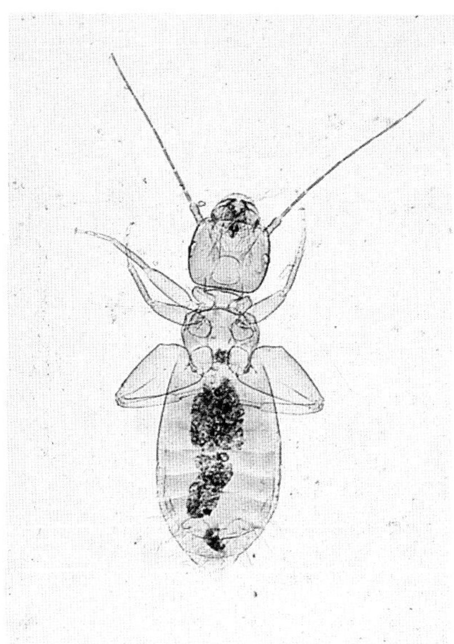

Figure 11.6 Psocoptera book louse, a cause of epidemics of infestation mistakenly ascribed to head lice. The environmental source must be identified and treated

Figure 11.7 *Columbicola columbae* (pigeon louse)

nits are found attached to the proximal portion of the hair shaft, especially in the retroauricular scalp (Figure 11.3).

Pediculus humanus corporis (body louse)

Body lice infest clothing rather than hair or skin, and leave the seams of clothing to obtain a blood meal. Body louse nits are also found in clothing seams. Although the adult body louse resembles a head louse (Figure 11.4), the two are easily distinguishable by the location of the nits (on hairs with head lice *vs* on clothing fibers with body lice). Clinically, a chronic case of body louse infestation presents as a hyperpigmented, lichenified, eczematous eruption (vagabond's disease). Body lice are implicated as the vector of epidemic typhus, trench fever and louse-borne relapsing fever.

Zoonotic lice

Other lice, such as Psocoptera, may infest houses

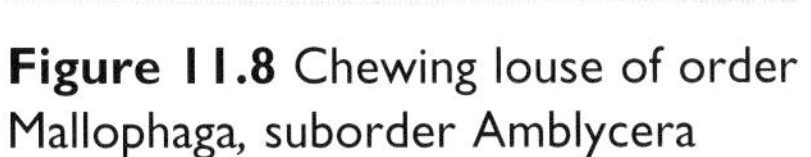

Figure 11.8 Chewing louse of order Mallophaga, suborder Amblycera

Figure 11.9 Another chewing louse of the order Mallophaga

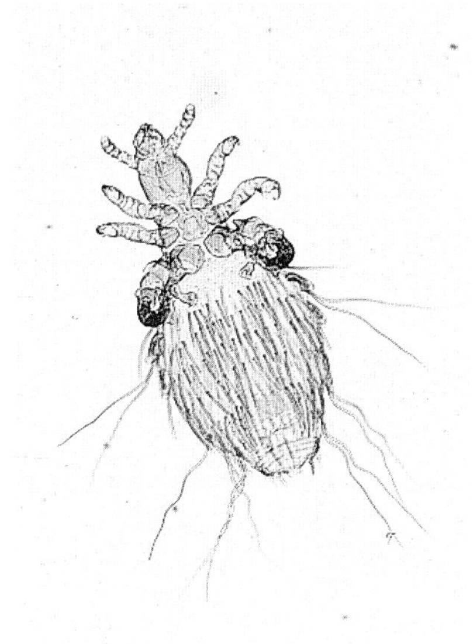

Figure 11.10 *Spermophilus mexicanus*, a sucking louse

or schools and cause scalp infestations (Figures 11.5–11.10). Reinfestation of the scalp will occur unless the environment is treated[3]. The gross morphological features of zoonotic lice are rather different from those of human lice.

In general, lice demonstrate marked host-specificity. Although widespread human infestation with 'non-human' lice may occur, it is uncommon.

Coleoptera (beetles)

Blister beetles (Figure 11.11) cause vesicular and bullous reactions. Hospital wards have experienced epidemics of bullous dermatitis caused by blister beetles which entered through open windows. Cantharidin, the vesicant, has long been used as a therapy for warts and molluscum contagiosum.

Figure 11.11 Blister beetles. Beetles come in a wide assortment of colors

Diptera (flies)

Fly larvae (maggots) cause human myiasis. *Cordylobia* flies (tumbu flies; Figure 11.12) are prevalent in Africa whereas *Dermatobia hominis* causes disease in Central and South America, and South Texas. *Cuterebra* flies occasionally produce human myiasis throughout the USA[4] which presents as deep furuncle-like lesions.

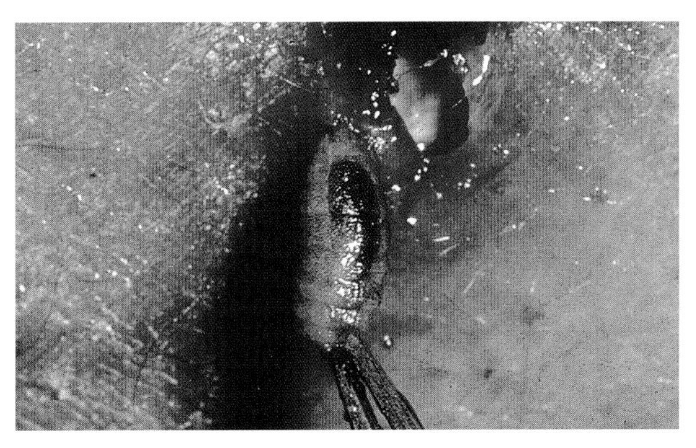

Figure 11.12 *Cordylobia anthropophaga* or tumbu fly maggot, a cause of myiasis. Courtesy of Dr Curt P. Samlaska, Las Vegas, NV

Figure 11.13 Psychodidae sandfly

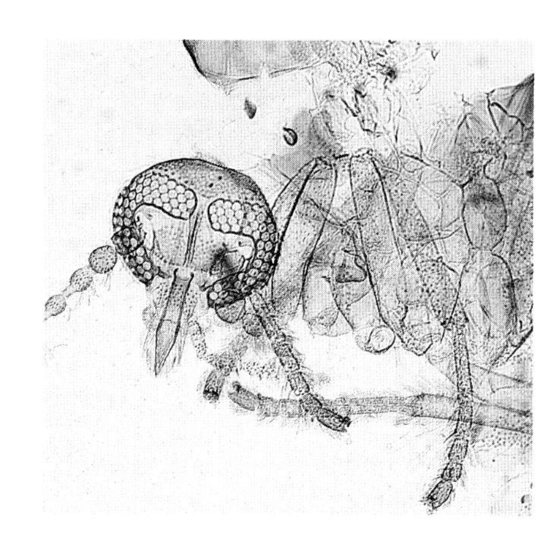

Figure 11.14 Psychodidae sandfly (higher magnification)

Flies frequently serve as disease vectors. Tularemia and loiasis (*Loa loa*) are transmitted by *Chrysops* (Tabanidae; deer flies), bartonellosis is transmitted by Psychodidae sandflies (*Lutzomyia;* Figures 11.13 and 11.14), and leishmaniasis is carried by sandflies belonging to *Phlebotomus*, *Lutzomyia* and *Psychodopygus* genera. *Leishmania mexicana* has been isolated in *Lutzomyia* flies in South Texas[5]. *Trypanosoma gambiense* (transmitted by the tsetse fly) causes African sleeping sickness, and *Onchocerca volvulus* is transmitted by black flies (Simuliidae, buffalo gnats, turkey gnats) in rural Africa and South America. *Dipetalonema* (*Acanthocheilonema, Mansonella*) worms are transmitted by midges, and mosquitoes (Culicidae) carry many viral and helminthic pathogens.

Hemiptera (true bugs)

Bedbugs

Bedbugs are flat and broad, with a proboscis which points downwards (Figure 11.15). *Cimex lectularius* and *C. hemipterus* (Asia) are the most commonly found bedbugs. They live in cracks and crevices, such as behind peeling wallpaper, and descend on their sleeping victims at night to obtain a blood meal. The bug often inflicts several bites in a row, taking a blood meal at each site. This linear array of papules or wheals has been termed 'breakfast, lunch and dinner'.

Closely related bugs commonly infest bird and bat dwellings. When the natural hosts migrate south, the bugs are forced to look for alternative sources of food.

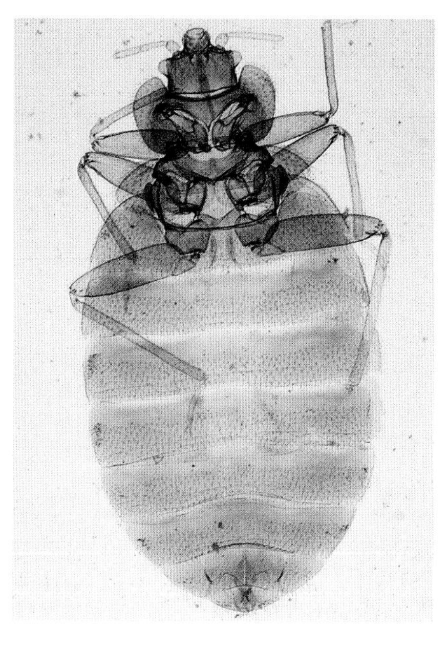

Figure 11.15 *Cimex* bedbug

Reduviid bugs

Kissing bugs (assassin bugs, cone-nosed bugs) of the order Reduviidae are common in the Americas. They are a cause of painful bites in many parts of the USA. *Triatoma* bugs (Figure 11.16) serve as a vector for American trypanosomiasis, which may present as an oculoglandular syndrome (Romaña's sign).

Wheel bugs

Wheel bugs (Figure 11.17) are commonly classified with the Reduviidae. They are easily recognizable by the spoked 'wheel' that protrudes from their dorsal surface, and are capable of delivering painful bites.

Figure 11.16 *Triatoma*, a Reduviid bug

Figure 11.17 Wheel bug

Figure 11.18 Fire ant

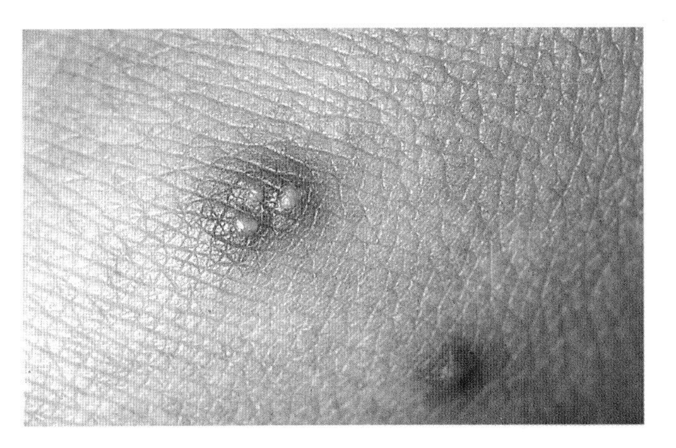

Figure 11.19 Characteristic grouping of fire ant stings

Hymenoptera

Fire ants (Solenopsis invicta and S. richteri)

These prolific mound-builders are now commonplace across the South and Southwest of the USA. Fire ants are rather small (Figure 11.18), but swarm in huge numbers when their mound is disturbed. The ants clamp onto the skin by their mandibles and rotate as they sting, producing a characteristic grouping or circle of stings (Figure 11.19). Acute pustular reactions are common. Delayed papulovesicular reactions may itch for weeks.

Bees and wasps

The manifestations of wasp and bee sting allergy are usually severe. Patients with a history of anaphylaxis from bee or vespid (Figure 11.20) stings should carry epinephrine.

Figure 11.20 Anaphylaxis to wasp stings should be treated promptly with epinephrine. Courtesy of Department of Preventive Health Services, US Army Medical Command Center and School, Fort Sam Houston, TX

Figure 11.21 Many less well-known caterpillars can cause epidemics of pruritus

Figure 11.22 *Megalopyge opercularis* (puss caterpillar)

Lepidoptera (moths and butterflies)

Most skin disease related to lepidopterids is caused by moths or moth caterpillars[6] (Figure 11.21). Skin lesions are due to urticating hairs on the body of the moth or caterpillar. Moths may be responsible for epidemics of pruritus. Gypsy moth caterpillars (*Lymantria dispar*, tent caterpillars) are a problem particularly in the northeastern and mid-Atlantic states[7].

Caripito itch, caused by contact with *Hylesia* moths, is common in South America[8], and ships docked at South American ports may become infested with the moths. Urticating hairs may be blown throughout the ship through the ventilation system, and the resulting epidemic of pruritus may last for weeks. American physicians practicing in port cities should consider moth dermatitis when faced with shipwide epidemics of 'scabies'.

Megalopyge opercularis (Figure 11.22), also known as the puss caterpillar, wooly asp, Italian asp, opossum bug and wooly slug, is particularly common in Texas, although it ranges from the mid-Atlantic seaboard to the southwestern states. The caterpillars live in tree foliage and it is not uncommon for them to fall onto someone pruning a tree. Contact with the caterpillar produces intensely painful, erythematous, edematous plaques.

Puss caterpillars are relatively short – around 35 mm

Figure 11.23 *Automeris io* (io moth caterpillar). Courtesy of Wilford Hall Medical Center, Department of the Air Force, Lackland AFB, TX

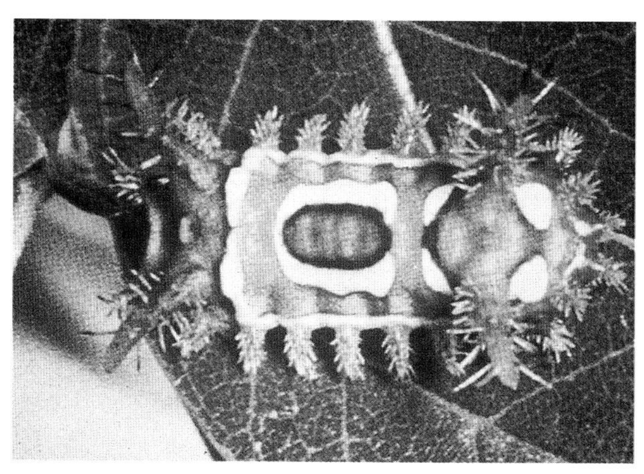

Figure 11.24 *Sibine stimulea* (saddleback caterpillar). Courtesy of Wilford Hall Medical Center, Department of the Air Force, Lackland AFB, TX

(1.5 inches) in length – and fat, and covered with a thick coat of light tan-to-brown hair. The hairs form a longitudinal midline ridge on the dorsal surface.

Automeris io, the io moth caterpillar (Figure 11.23), is found throughout the eastern and southern USA. The caterpillar is light green with red and white stripes.

Sibine stimulea, the saddleback caterpillar (Figure 11.24), is found in the southeastern states. The caterpillar has a distinctive shape, resembling a miniature child's ride-on toy, with colorful stripes running along its sides.

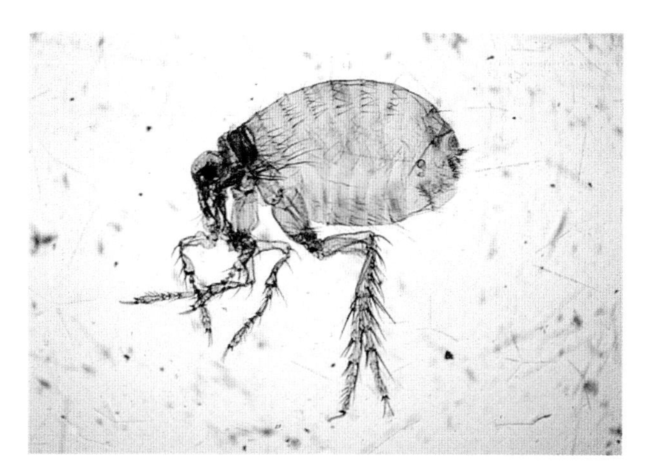

Figure 11.25 *Echidnophaga gallinacea* (stick-tight flea)

Siphonaptera (fleas)

Fleas are characterized by large powerful legs and a flat (side-to-side) body habitus (Figure 11.25). Common fleas include: *Pulex irritans*, the human flea (Figure 11.26); *Ctenocephalides felis*, the cat flea (Figure 11.27); and *C. canis*, the dog flea. Fleas are absolutely non-host-specific; cat fleas are the usual fleas found on dogs, which may also be infested by human fleas.

Flea bites are intensely pruritic and tend to be grouped on the lower legs or hands and forearms (after petting or handling an infested animal). It is common for one member of a family to manifest a far greater degree of flea-bite sensitivity compared with other family members.

Tunga penetrans, the chigoe flea or jigger, is found in sandy areas in South America, the West Indies and tropical Africa, and cases have been reported in the USA[9]. The gravid female flea burrows into an area of exposed skin (usually a toe) and, once embedded, increases to approximately the size of a small pea. The head of the flea lies within the dermis and the body is surrounded by acanthotic epidermis. Excision is the treatment of choice. The flea may first be killed by application of a chloroform- or ether-soaked pledget.

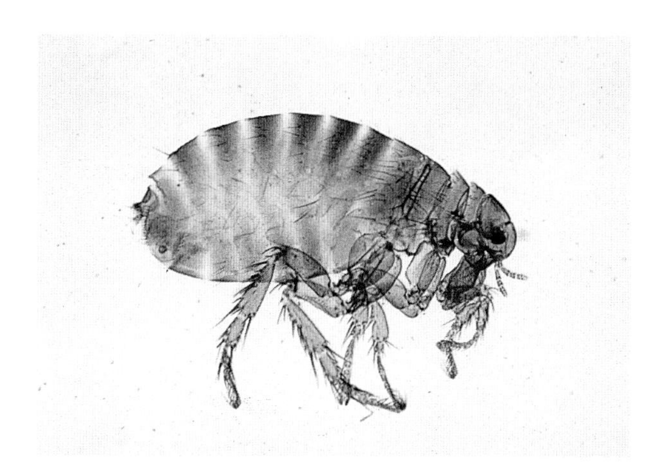

Figure 11.26 *Pulex irritans*: This 'human flea' has little host-specificity and, similar to *Ctenocephalides felis*, commonly infests dogs

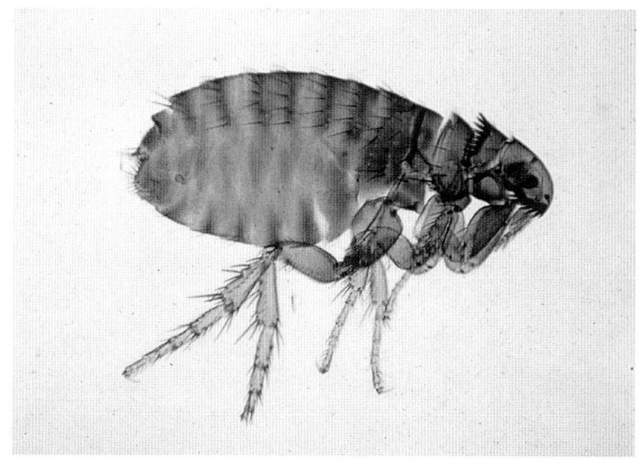

Figure 11.27 *Ctenocephalides felis* (cat flea) is a vector for endemic typhus and cat-scratch disease

Fleas are common vectors of disease. Plague, transmitted by the rat flea *Xenopsylla cheopis* (Figure 11.28), still occurs as an endemic zoonosis in the American southwest. Endemic typhus is also seen as a zoonosis in the southwestern states. The causative organism, *Rickettsia felis* (ELB agent), is carried by opossums and transmitted by the cat flea *C. felis*[10]. *Rickettsia typhi*, another agent of endemic typhus, is transmitted by *Xenopsylla cheopis* and *C. felis*, and cat-scratch disease may be transmitted by cat fleas[11,12].

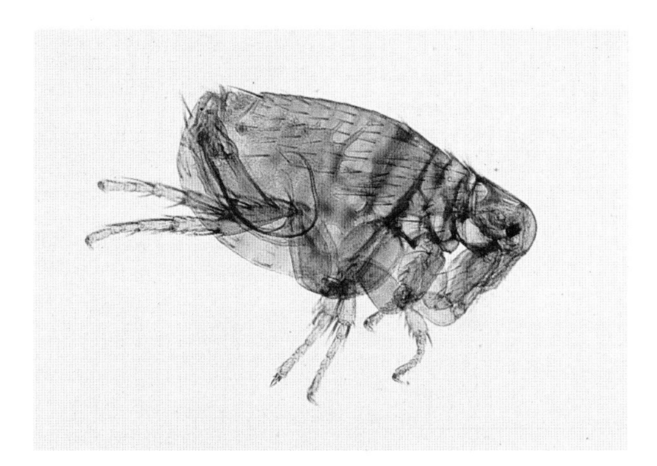

Figure 11.28 *Xenopsylla cheopis* is a vector for plague and endemic typhus

References

1. Signor RJ, Love J, Boucree MC. Scalp infection with *Phthirus pubis*. *Arch Dermatol* 1989;125:133

2. Ashkenazi I, Desatnik HR, Abraham FA. Yellow mercuric oxide: A treatment of choice for phthiriasis palpebrum. *Br J Ophthalmol* 1991;75:356–8

3. Coulthard M, Heaney J. Scalp infestation by *Liposcelis mendax*. *Br J Dermatol* 1991;125:400–1

4. Elgart ML. Flies and myiasis. *Dermatol Clin* 1990;8: 237–44

5. McHugh CP, Grogl M, Kreutzer RD. Isolation of *Leishmania mexicana* (Kinetoplastida: Trypanosomatidae) from *Lutzomyia anthophora* (Diptera: Psychodidae) collected in Texas. *J Med Entomol* 1993;30:631–4

6. Rosen T. Caterpillar dermatitis. *Dermatol Clin* 1990; 8:245–51

7. Aber R, DeMelfi T, Gill T, *et al.* Rash illness associated with gypsy moth caterpillars – Pennsylvania. *MMWR* 1982;31:169–70

8. Dinehart SM, Archer ME, Wolf JE, *et al.* Caripito itch: Dermatitis from contact with *Hylesia* moths. *J Am Acad Dermatol* 1985;13:743–7

9. Sanus ID, Brown EB, Shepard TG, *et al.* Tungiasis: Report of one case and review of the 14 reported cases in the United States. *J Am Acad Dermatol* 1989;20:941–4

10. Schriefer ME, Sacci JB, Taylor JP, *et al.* Murine typhus: Updated roles of multiple urban components and a second typhus-like rickettsia. *J Med Entomol* 1994; 31:681–5

11. Higgins JA, Radulovic S, Jaworski DC, *et al.* Acquisition of cat scratch disease agent *Bartonella henselae* by cat fleas (Siphonaptera: Pulicidae). *J Med Entomol* 1996;33:490–5

12. Zangwill KM, Hamilton DH, Perkins BA, *et al.* Cat scratch disease in Connecticut. *N Engl J Med* 1993; 329:8–13

Chapter 12 Helminthic pathogens (worms)

Enterobius vermicularis (pinworm)

Microscopic preparations may be useful in the evaluation of perianal pruritus. Pinworms migrate to the anus during the night to lay eggs. Clear adhesive tape preparations are easily obtained from perianal skin. They are best gathered early in the morning, before the patient has bathed or passed a bowel movement. The adhesive tape is applied repeatedly to the skin; the tape is then mounted like a coverslip onto a glass microscopy slide. As the eggs are found on skin and not in the stool, adhesive tape preparations are clearly superior to stool specimens.

Eggs are embryonated and have a thick hyaline shell, which resembles a blunt rugby ball in shape and is characteristically flattened on one side (Figure 12.1). The female pinworm migrates onto the anal mucosa at night to lay eggs. Occasionally, the adult worm, rather than the eggs, is picked up by the adhesive tape preparation. Adult female pinworms have a blunted 'nose', pointy tail and a constriction followed by a bulbous dilatation of the esophagus (Figures 12.2 and 12.3).

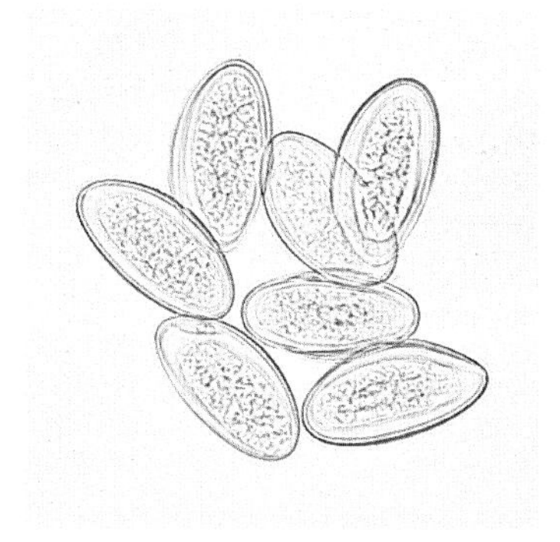

Figure 12.1 *Enterobius vermicularis* (pinworm) ova have a flattened rugby ball-like shape

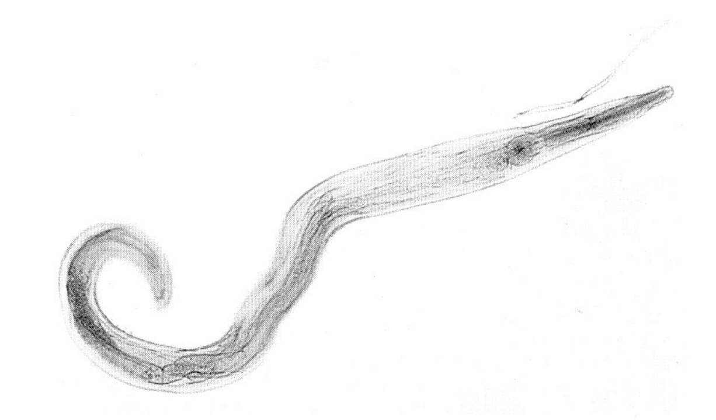

Figure 12.2 Adult female pinworm has a bulbous dilatation of the esophagus

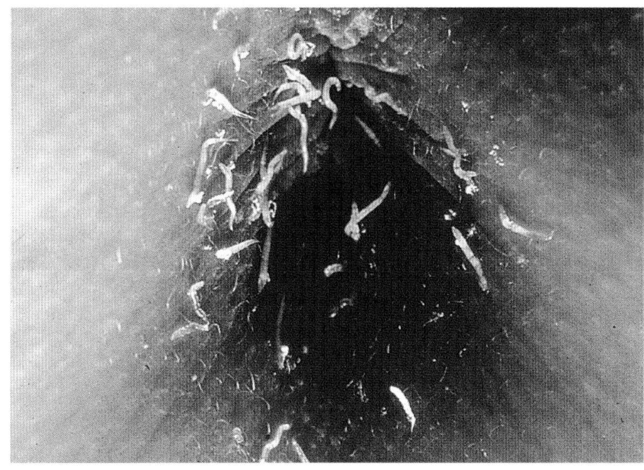

Figure 12.3 Adult pinworms. Courtesy of Dr Martin Weber, with permission of the *New England Journal of Medicine*

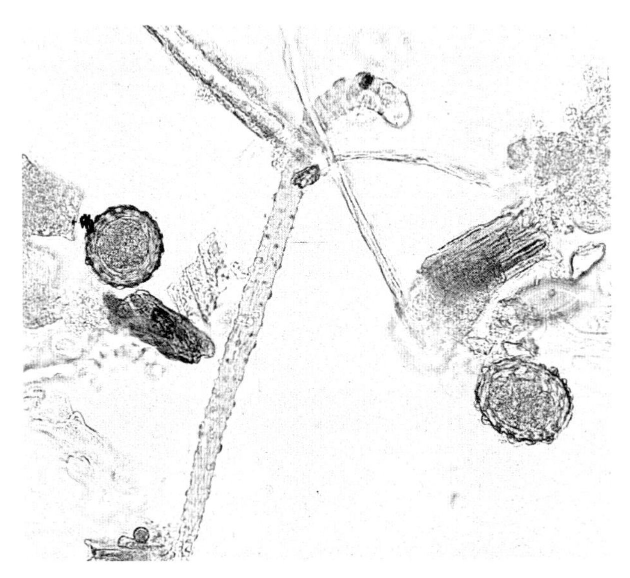

Figure 12.4 *Ascaris lumbricoides* ova are brown (bile-stained) and resemble a fluted pie shell

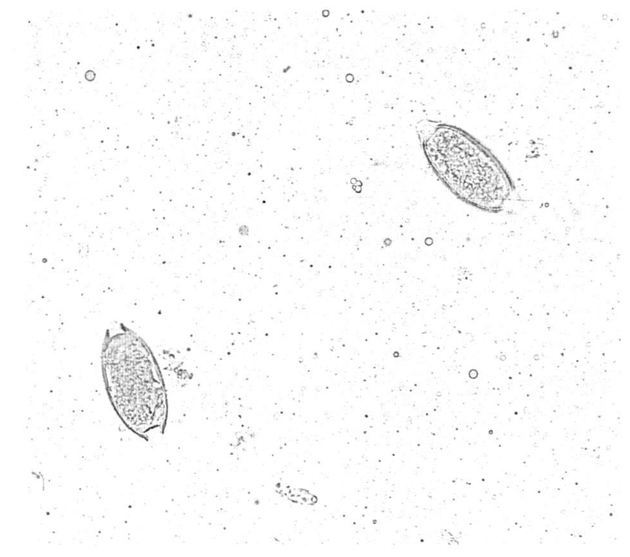

Figure 12.5 *Trichuris trichiura* ova are brown (bile-stained) with 'plugs' at both ends

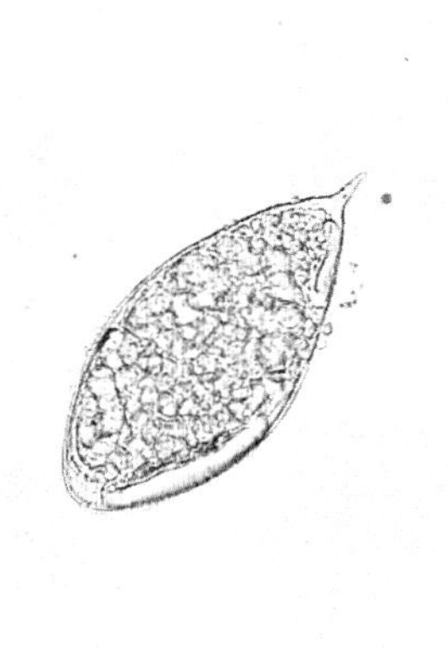

Figure 12.6 *Schistosoma hematobium* ova are typically oval in shape with a thin apical spine

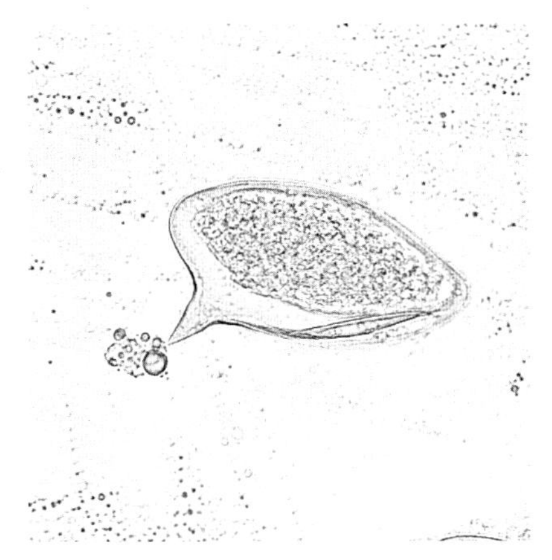

Figure 12.7 *Schistosoma mansoni* ova have a characteristic oval shape and a thick lateral spine

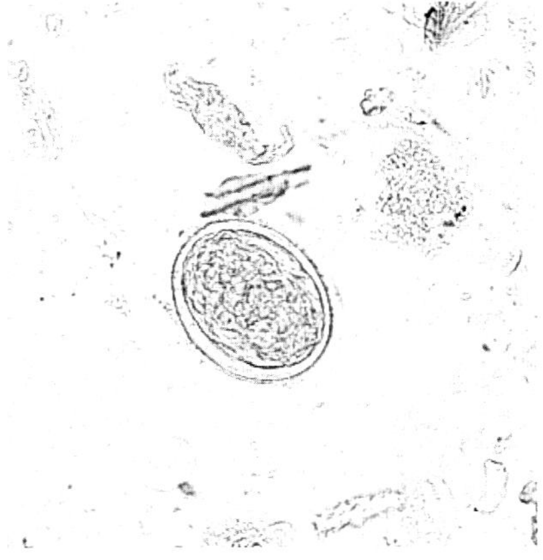

Figure 12.8 *Schistosoma japonicum* ova are rounder with an inconspicuous to absent spine

Other intestinal parasites

Occasionally, the eggs of other gastrointestinal pathogens are found on perianal skin (Figures 12.4 and 12.5). Some gastrointestinal ova and parasites are easily recognized but, if a clinician has any doubts regarding identification, the specimen in question should be forwarded to a large clinical laboratory for confirmation.

Parasite ova may be infective and, therefore, should be handled and disposed of appropriately.

Schistosoma ova

Schistosomal eggs may occasionally produce papillomatous or verrucous skin lesions. The ova of *Schistosoma hematobium* are characterized by an oval shape and a thin apical spine (Figure 12.6) whereas those of *S. mansoni* are oval with a thick lateral spine (Figure 12.7). *Schistosoma japonicum* produces rounder, smaller eggs (Figure 12.8), with a spine that may be inconspicuous to absent.

Section III Fungi

by Dennis E. Babel

Chapter 13	Laboratory methods in mycology	68
Chapter 14	Dermatophytes	71
Chapter 15	Non-dermatophytic hyaline and dematiaceous molds	86
Chapter 16	Yeasts	98
Chapter 17	Dimorphic fungal pathogens	103

Chapter 13 Laboratory methods in mycology

Fungi are ubiquitous eukaryotic lifeforms and are taxonomically placed in the kingdom Fungi. This kingdom is divided into four major phyla, according to the characteristics of organism reproduction. Most of the fungal species known to cause infections in humans belong to the phylum Deuteromycetes. These fungi usually reproduce asexually and present with a yeast and / or mold morphology.

Yeast morphology

A fungus with a yeast morphology reproduces through budding or blastogenesis. The parent cell may give rise to single or multiple daughter cells (blastoconidia). The size, shape, arrangement, pigmentation and possible encapsulation of blastoconidia accord them characteristics that may sometimes lead to their successful identification *in vivo*. This aspect is especially important in the diagnosis of systemic mycoses as it allows the physician to make a presumptive pathogen identification weeks before the organism is isolated and identified by traditional fungal culture methods.

Mold morphology

A fungus growing in a mold morphology presents as a filamentous structure composed of two parallel extending walls. These filaments (hyphae) may branch, may produce crosswalls called 'septations' or remain aseptate (cenocytic). Some molds (as with some species of yeast) are capable of producing melanin and are referred to as 'dematiaceous'. In tissue, dematiaceous fungi may be gold to dark brown in color because of this attribute. Considerations of diameter, branching angle, septations, pigment production and unique growth patterns help to categorize molds.

Dimorphicity

Many of the fungi capable of causing disease in humans are seen only with a yeast or mold morphology and are referred to as 'monomorphic'. Other species, including some of those responsible for the systemic mycoses, grow in more than one form. Most frequently, these include organisms capable of reproducing as both a yeast and a mold. Fungi which are 'thermodimorphic' are able to grow in a hyphal morphology when cultured at 25°C (room temperature) and in the yeast phase when cultured at 37°C (body temperature).

Occasionally, however, a totally different morphology is observed in tissue and represents neither a true yeast nor mold, for example, the spherules seen in coccidioidomycosis or the 'sclerotic bodies' characteristic of chromoblastomycosis. The various characteristic fungal morphologies seen *in vivo* are summarized in Table 13.1.

Direct microscopy

Familiarity with these unique fungal forms as seen in patient specimens is essential for early diagnosis. Observations may be made by direct micros-

Table 13.1 Fungal morphology *in vivo*

Hyphal morphology	*Indeterminant morphology*	*Yeast morphology*
Cenocytic hyphae (e.g. mucormycosis)	Sclerotic bodies (e.g. chromoblastomycosis)	Blastoconidia (e.g. Paracoccidioidomycosis)
Septate hyphae (e.g. dermatophytosis)	Spherules with endospores (e.g. coccidioidomycosis)	Blastoconidia (in chains) (e.g. lobomycosis)
Sulfur grains (e.g. eumycotic mycetoma)	Sporangium with spores (e.g. rhinosporidiosis)	Pseudohyphae (e.g. candidiasis)

copy of infected tissue samples or lesion exudate. A number of different clearing solutions may be applied to patient material to assist in direct microscopy (Table 13.2). These agents help to distribute the specimen to facilitate visualization of any fungal structures that may be present, and include simple saline and potassium or sodium hydroxide in various formulations as well as preparations which incorporate various coloring agents. Microbiological stains (Table 13.3) may also be used to detect the presence and define the morphology of fungi in exudate smears and touch imprints of biopsy specimens.

Table 13.2 Clearing solutions for fungal microscopy

Solution	*Property*
Saline	distributes material; no coloring
KOH 10%	clears specimen; no coloring
KOH 20% with DMSO	clears specimen rapidly; no coloring
KOH with Parker ink	clears specimen; colors light violet
Chlorazol black E	clears specimen; colors blue-black
KOH with calcofluor	clears specimen; fluoresces yellow

KOH, potassium hydroxide; DMSO, dimethyl sulfoxide

Histopathology

Biopsy specimens obtained from fungal lesions may reveal the morphology of the infectious agent

Table 13.3 Fungal staining agents

Stain	*Property*
Gram's	blue / black (most fungi)
Giemsa	blue / black (most fungi)
H & E	pale blue (many fungi)
PAS	pink / red (most fungi)
Silver	blue / black (most fungi)
Masson–Fontana	melanin appears brown (most dematiaceous fungi)
Mucicarmine	pink / red (yeast capsule)
Alcian blue	blue (yeast capsule)
India ink preparation	darkens field, highlights capsule
Acid-fast	pink / red (filamentous bacteria, such as *Nocardia* spp)

H & E, hematoxylin and eosin; PAS, periodic acid–Schiff

in vivo as well as the host response to its invasive presence. The appropriately stained histopathological section may provide the clinician with proof of the presence of a fungal pathogen, clues to its identity, the extent of infection and the patient's ability to respond to this invasion.

Culture

The ultimate identification of fungal isolates requires a review of their characteristics *in vitro*, including both morphological and physiological characteristics. Morphological considerations (such as gross and microscopic fungal colony features) may be directly affected by the culture system used (Table 13.4). Some media restrict the growth of contaminants to facilitate isolation of the fungal pathogens. Other agars are formulated to enhance the sporulating capabilities or pigment production of certain species. Physiological characteristics include nutritional requirements, temperature tolerance, enzymatic activity and carbohydrate assimilation or fermentation. These various testing systems are further described in Chapter 14 (see pages 75–85).

Table 13.4 Fungal culture media

Media	*Attributes*
Sabouraud's dextrose agar (SDA)	supports growth of most molds and yeasts; shows typical morphology
SDA with chloramphenicol and cyclohexamide (SDA with C & C)	isolation medium; inhibits most bacterial and fungal contaminants
Brain–heart infusion agar (BHIA)	enriched medium to support growth of the more fastidious fungi
BHIA with blood (antibiotics may be added to inhibit bacterial contaminants)	converts dimorphic fungi from mold to yeast phase
Cornmeal agar with Tween 80 (chlamydospore agar)	stimulates chlamydospore production in *Candida albicans*
Potato dextrose agar (PDA)	enhances typical pigmentation of some dermatophytes
Potato flake agar (PFA)	enhances typical sporulation of many molds
Trichophyton agars (1–7)	ascertains characteristic nutritional requirements of various dermatophyte species
Christensen's urea agar	detects urease activity of certain fungi
Rice grain ('polished rice') medium	helps to differentiate *Microsporum canis* from *M. audouinii*

Chapter 14 Dermatophytes

Pathogenic species

The dermatophytes are a group of related monomorphic molds comprising more than forty species placed into the genera *Trichophyton*, *Microsporum* or *Epidermophyton*. These fungi are capable of deriving their total nutrition from keratin by elaborating various keratinase enzymes. Many species are nonpathogenic and grow only in the soil. Around a dozen species are responsible for the majority of human dermatophytoses found worldwide.

Epidemiology

There are three natural reservoirs of dermatophytes:
(1) Geophilic;
(2) Zoophilic; and
(3) Anthropophilic.

Early dermatophyte species originated in the soil where they probably subsisted on keratin that was dropped by passing animals. Over time, some of these species evolved to assume a different environmental niche, which now includes both animal and human reservoirs.

Pathogens which are geophilic reside in the soil and initiate clinical disease when a subject comes into contact with this contaminated substrate. The most common example of a dermatophyte species from this source is *Microsporum gypseum*. Those species which have evolved to become obligate parasites of certain animals are classified as zoophilic and normally reproduce only on animal hosts. Human infection by these species usually occurs after direct contact with the animal reservoir. Transmission of a zoophilic pathogen from an initially infected human to a second human host occurs less frequently, and is of limited virulence and disease duration. Most human cases of tinea are attributable to anthropophilic species (Table 14.1). These obligate parasites of humans

Table 14.1 Ecology of dermatophytes reported to cause tinea in North America

Anthropophilic	*Zoophilic*	*Geophilic*
Trichophyton rubrum	*Trichophyton verrucosum*	*Microsporum gypseum*
Trichophyton mentagrophytes var. *interdigitale*	*Trichophyton mentagrophytes* var. *mentagrophytes*	*Microsporum fulvum*
Trichophyton tonsurans	*Trichophyton equinum*	
Trichophyton schoenleinii	*Microsporum canis*	
Microsporum audouinii	*Microsporum nanum*	
Trichophyton ferrugineum	*Trichophyton gallinae*	
Epidermophyton floccosum		

cannot reproduce in the soil and infections are difficult to establish in animal models. When the clinician is faced with the problem of repeated reinfection in a previously cured patient, consideration must be given to the epidemiology and environmental source involved.

Clinical disease

A dermatophytosis or tinea is an infection of the dead, keratinized tissues of the body, including hair, skin and nail. Invasion of viable tissue is uncommon, and dissemination from a cutaneous site to internal organs is extremely rare. The specific form of clinical infection acquires its name according to the anatomical site involved (for example, infection of the foot is called tinea pedis, and of the scalp, tinea capitis).

Cutaneous lesions may be annular, serpiginous, eczematous, erythematous, violaceous, papular, pustular, follicular, nodular, granulomatous, vesicular to bullous or present as areas of non-inflammatory alopecia.

Tinea capitis may be non-fluorescent, as when due to *Trichophyton tonsurans*, or may demonstrate a bright blue-green color under ultraviolet fluorescent light due to the presence of pteridine, which is produced by dermatophytes such as *Microsporum canis*. Support for a suspected clinical diagnosis may be readily obtained by direct microscopy, but the ultimate pathogen identification is achieved by fungal culture. Consideration should be given to the best means to acquire a patient specimen to maximize the chances of a successful result with these procedures (Table 14.2).

Morphology *in vivo*

Direct microscopic examination using potassium hydroxide (KOH) of material taken from the site of infection demonstrates hyaline, septate hyphae and occasional arthroconidia in skin and nail specimens. Infected hairs may reveal hyphae separating into masses of large (5–8 μ in diameter) arthroconidia within the shaft (endothrix arthroconidiation) or masses of small arthroconidia (2–4 μ in diameter) on the shaft surface (ectothrix arthroconidiation).

Morphology *in vitro*

Primary isolation of the infecting dermatophyte species is best performed on a restrictive medium such as Sabouraud's dextrose agar (SDA) which may incorporate chloramphenicol to inhibit bac-

Table 14.2 Specimen collection from cases of dermatophytoses

Tissue	*Collection technique*	*Unacceptable specimen*
Hair	clean area of alopecia; scrape black dots and scale; pluck fluorescent hairs	long strands of hair; hair clippings
Skin (scaling lesions)	clean lesion with alcohol; scrape scale from advancing edge	scrapings from central clearing area of lesion; moist, macerated, intertriginous tissue
Skin (vesicular lesions)	clean lesion with alcohol; excise portion of vesicle roof	scrapings from vesicle base; vesicular fluid
Nail (superficial; white spot)	clean nail with alcohol; scrape white chalky surface	nail clippings
Nail (distal subungual)	trim back plate to advancing edge of mycosis and discard; collect specimen from mycotic edge closest to proximal nailfold	distal subungual debris; nail clippings; whole removed nail
Nail (proximal subungual)	collect specimen from mycotic edge most distal to cuticle	proximal subungual debris; nail clippings

Table 14.3 Sporulation characteristics of dermatophyte genera

	Trichophyton	*Microsporum*	*Epidermophyton*
Macroconidia			
Occurrence	rare	many	many
Shape	cigar / pencil	spindle / tapered	club / blunt
Wall	thin / smooth	thick / echinulate	thin / smooth
Microconidia			
Occurrence	numerous	occasional	not seen
Shape	species-specific	non-specific	–

terial contaminants and cyclohexamide to inhibit fungal saprophytes. Characteristic colony morphology should include consideration of texture and topography as well as both obverse and reverse pigmentation. A lactophenol cotton blue (LPCB) microscopy preparation made from fungal colony material should be examined for characteristic reproductive forms, including microconidia and macroconidia. The dermatophytes as a group are categorized into different genera and species on the basis of this asexual sporulation (Table 14.3). Observation of unique hyphal features, such as chlamydoconidia or favic chandelier formation, may provide additional clues to identification.

Other laboratory procedures which may be used to identify dermatophytes include:

- *Trichophyton* agars: Subculture of the primary isolate to one of these secondary medias may reveal the pathogen's unique nutritional requirement (Table 14.4);

- Starch agars: Subculture to a starch agar such as potato dextrose agar (PDA) or cornmeal dextrose agar (CMA) may induce characteristic pigmentation: For example, *Trichophyton rubrum* is cherry-red; *T. mentagrophytes* has no pigmentation; *Microsporum canis* is lemon-yellow; and *M. audouinii* is salmon-pink;

Table 14.4 Dermatophytes with nutritional requirements

Trichophyton tonsurans	thiamine
Trichophyton violaceum	thiamine
Trichophyton megnini	L-histidine
Trichophyton equinum	niacin
Trichophyton verrucosum	thiamine + inositol

- Urease test: Subculture of an isolate to Christensen's urea agar demonstrates fungal urease enzyme activity by changing the color of the medium to red. Thus, *T. mentagrophytes* is urease-positive whereas *T. rubrum* is urease-negative;

- Hair perforation test: This *in vitro* Petri dish test uses sterilized hair clippings as a keratin substrate and determines possession of a hyphal appendage, the so-called eroding frond, unique to certain species of dermatophyte. Thus, *T. mentagrophytes* is hair perforation test-positive whereas *T. rubrum* does not perforate hair and is test-negative;

- Rice grain test: This procedure determines fungal ability or inability to grow on a substrate of polished rice. Thus, *M. canis* is rice grain-positive whereas *M. audouinii* is rice grain-negative (does not grow on polished rice).

Trichophyton rubrum

Source Anthropophilic (worldwide)

Clinical features Infects hair, skin and nails (one hand–two foot syndrome, Majocchi's granuloma)

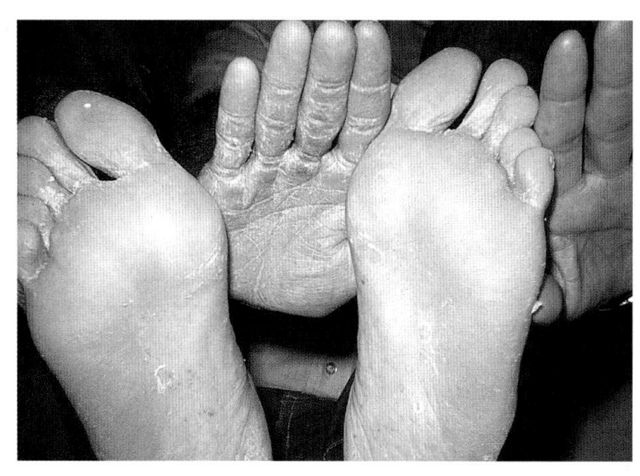

Figure 14.1 One-hand–two-foot syndrome, a diffuse scaly tinea of one palmar and both plantar surfaces, is most frequently caused by *T. rubrum*

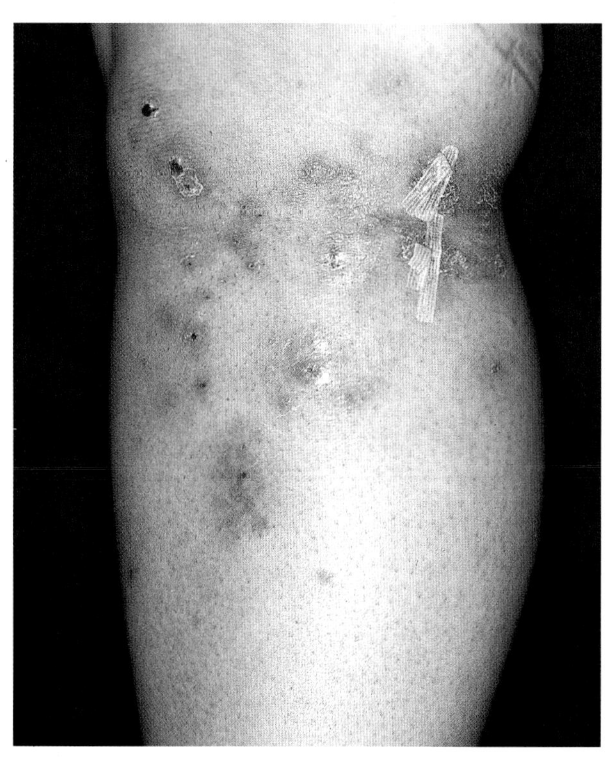

Figure 14.2 Majocchi's (*T. rubrum*) granuloma presents as erythematous to violaceous scaling nodules in response to deep hair follicle invasion by a dermatophyte

In vivo Hyaline septate hyphae

In vitro

Colony (macroscopically)

Topography: mounded center (umbonate), flattened periphery

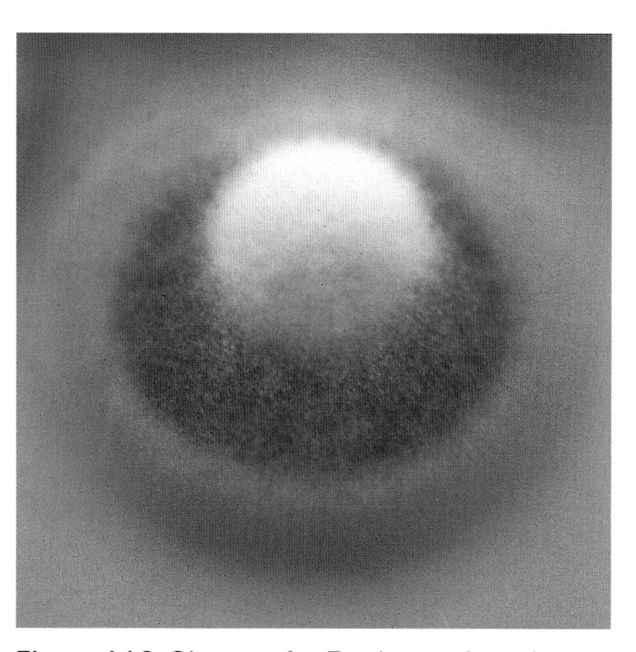

Figure 14.3 Obverse of a *T. rubrum* colony shows a white, cottony mound-like (umbonate) center and a flattened periphery

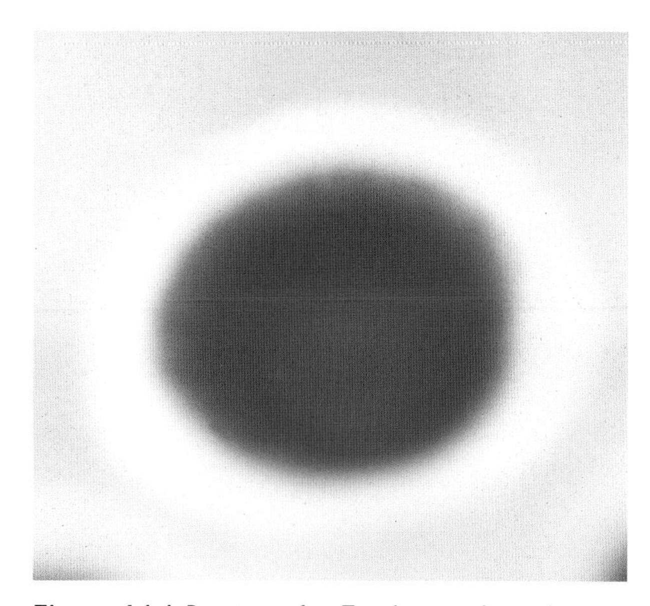

Figure 14.4 Reverse of a *T. rubrum* colony shows a dark-brown, non-diffusing pigment

Texture: cottony to powdery (rare)

Pigmentation: obverse is white in the center, yellow to reddish-brown at the periphery; reverse is reddish-brown, rarely non-pigmented

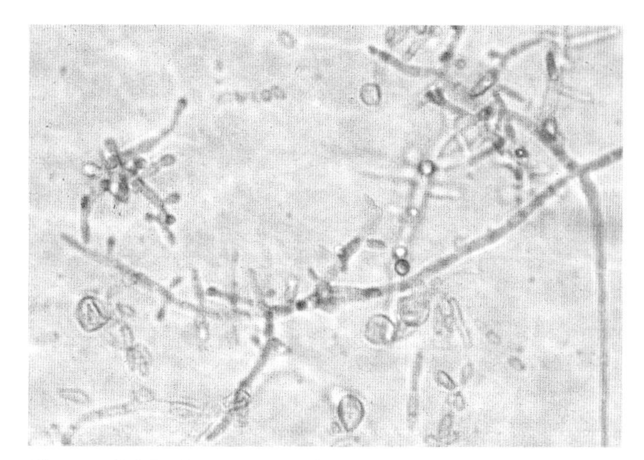

Figure 14.5 Microscopy of *T. rubrum* shows single club/ tear-shaped microconidia attached directly to hyphae (LPCB)

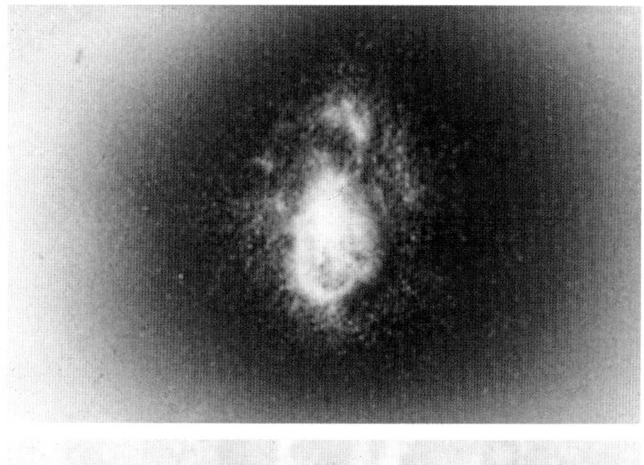

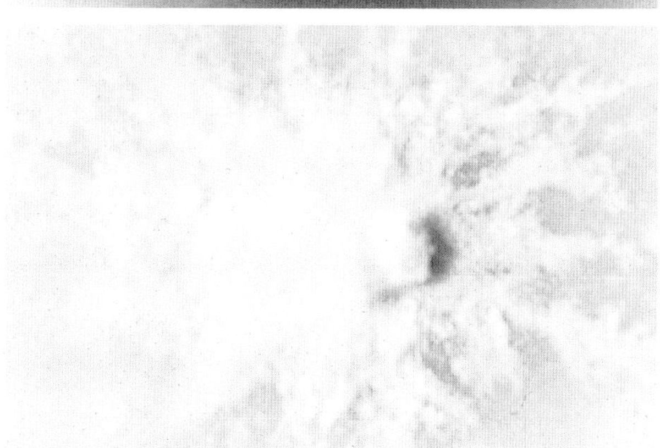

Figure 14.6 On PDA, *T. rubrum* (upper) produces a red pigment, which differentiates it from *T. mentagrophytes* (lower)

Colony (microscopically)

Hyphae: regular

Macroconidia: rarely seen; thin, smooth-walled, pencil-shaped; 5–15 segments

Microconidia: few to many; peg- or pyriform- or tear-shaped; single

Special features

Starch agars: cherry-red pigment is produced on either PDA or CMA

Urease test: negative

Hair perforation: negative

Trichophyton mentagrophytes

Source Anthropophilic (var. *interdigitale*); zoophilic (var. *mentagrophytes* and others) worldwide; zoophilic strains are contracted from many animal sources, including mice, rats, guinea pigs, rabbits, birds, dogs and horses

Clinical features Infects hair, skin and nails (leukonychia mycotica, vesicular tinea pedis)

In vivo Hyaline septate hyphae

In vitro

Colony (macroscopically)

Topography: mound-like (var. *interdigitale*); flat (var. *mentagrophytes*)

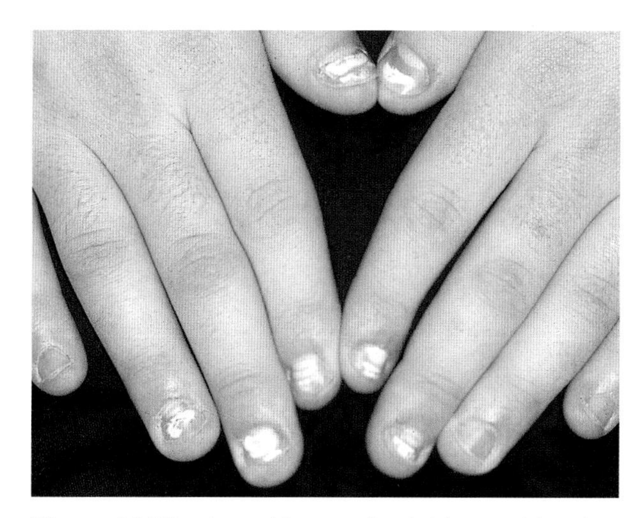

Figure 14.7 Leukonychia mycotica (white spot) is a dermatophyte infection of the top of the nail plate, resulting in chalky erosions, but no build-up of subungual debris

Texture: cottony (var. *interdigitale*); granular (var. *mentagrophytes*)

Pigmentation: obverse is white (var. *interdigitale*) or cream (var. *mentagrophytes*); reverse is non-pigmented; red / brown strains uncommon

Colony (microscopically)

Hyphae: regular with occasional spiral coils

Macroconidia: rare; thin, smooth-walled, club-shaped; 5–6 segments

Microconidia: numerous round, grape-like clusters, a few tear-shapes

Special features

Starch agars: no pigment on either PDA or CMA

Urease test: positive

Hair perforation: positive

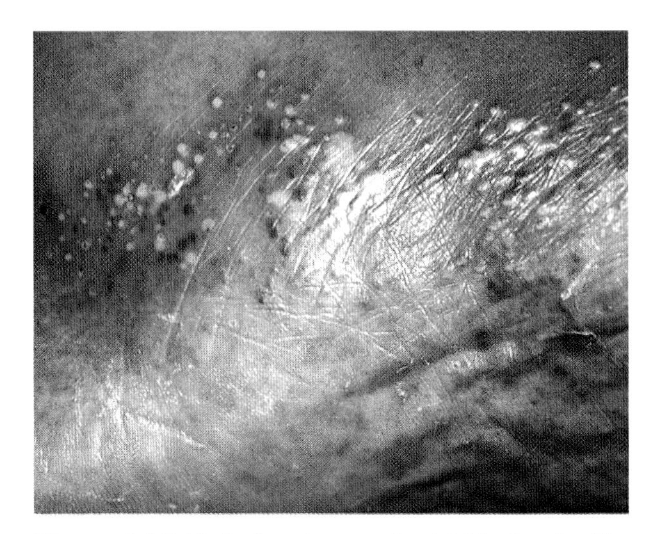

Figure 14.8 Vesicular tinea pedis. Initial pinpoint-like vesicles enlarge and may become multiloculated and, occasionally, form large, tense, fluid-filled bullae

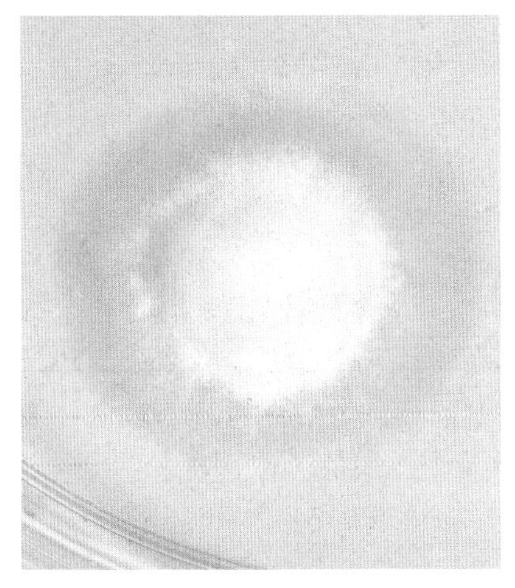

Figure 14.9 Obverse of a *T. mentagrophytes* var. *interdigitale* colony presents as a white, cottony mound with a discrete edge

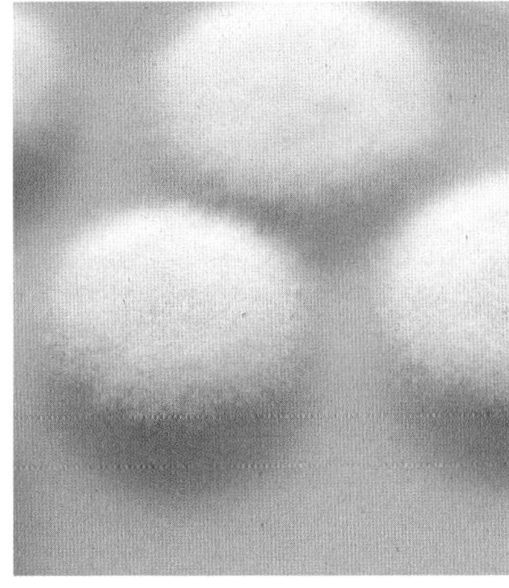

Figure 14.10 Obverse of a *T. mentagrophytes* var. *mentagrophytes* colony shows a flat, granular, cream-colored center with a cottony, white, rolled edge

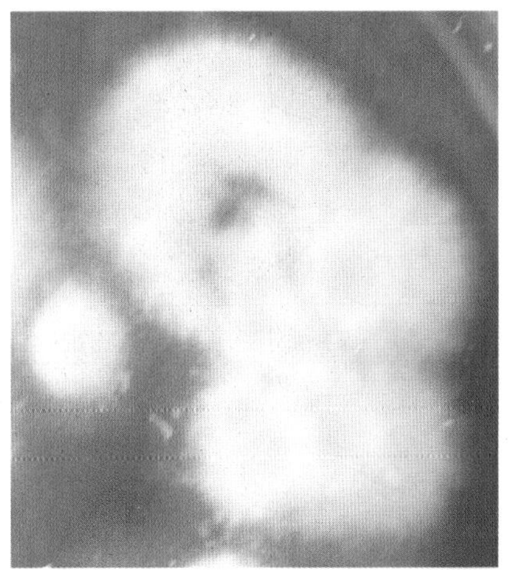

Figure 14.11 Reverse of a *T. mentagrophytes* colony is non-pigmented to occasionally light yellow-brown

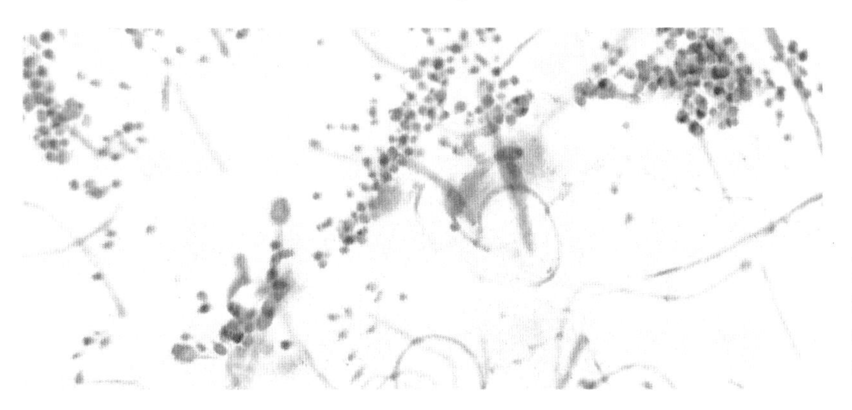

Figure 14.12 Microscopy of *T. mentagrophytes* reveals spiral coils and numerous clusters of round microconidia (LPCB)

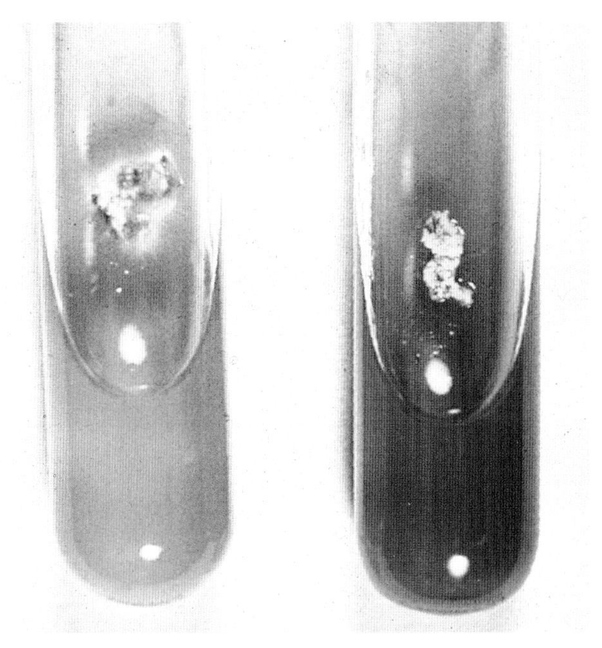

Figure 14.13 Urease test with *T. rubrum* (left) and *T. mentagrophytes* (right) shows negative and positive (red color development) results, respectively

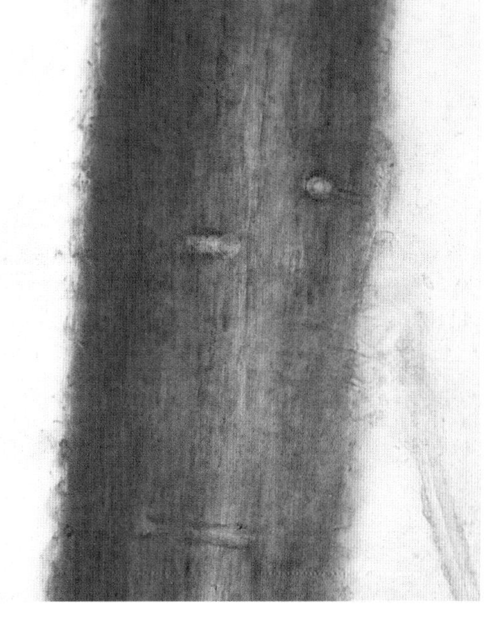

Figure 14.14 Hair perforation test. This hair fragment shows many clear erosions, indicative of (*T. mentagrophytes)* positive perforation (LPCB)

Trichophyton tonsurans

Source Anthropophilic (worldwide, primarily found in Central and North America)

Clinical features Infects hair, skin and nails (non-fluorescent; endothrix black-dot tinea capitis)

In vivo Hyaline septate hyphae (5–8 μ endothrix arthroconidia in hair)

In vitro

Colony (macroscopically)

Topography: flat, plicate, cerebriform or crateriform

Texture: powdery or suede-like with a feathery edge

Pigmentation: obverse is white, tan, yellow or brown; reverse is mahogany-red to dark brown

Colony (microscopically)

Hyphae: variable in shape ('caterpillar' hyphae); chlamydoconidia

Macroconidia: rare; medium-sized, smooth-walled, cigar-shaped; 4–6 segments

Microconidia: many peg-, round, tear- or balloon-shaped forms; sessile or stalked

Special features

Nutritional test: thiamine requirement

Urease test: positive

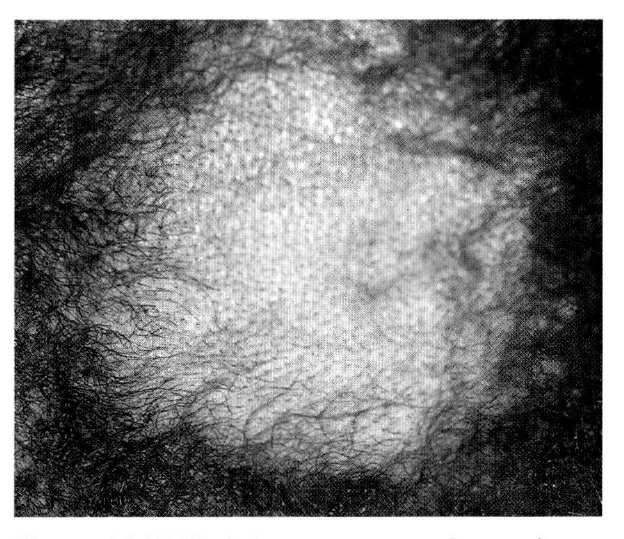

Figure 14.15 Black-dot tinea capitis. Areas of non-inflammatory alopecia show discrete black dots, indicating infected hairs broken off at the follicular orifice

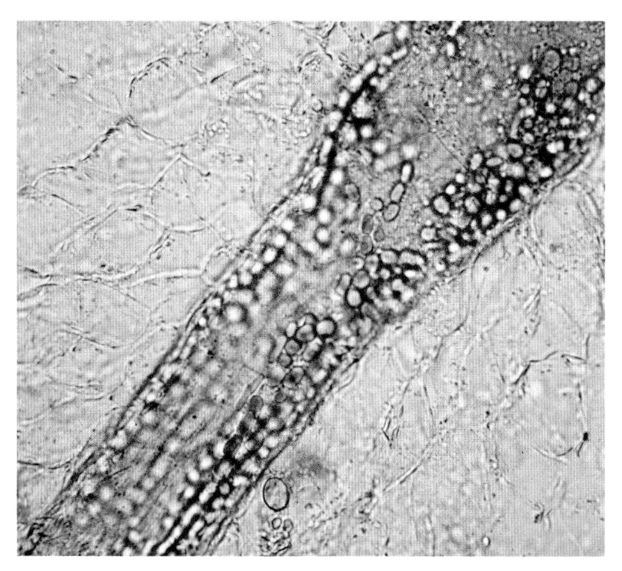

Figure 14.16 Endothrix tinea capitis. This endothrix-infected hair fragment is packed with arthroconidia. (KOH)

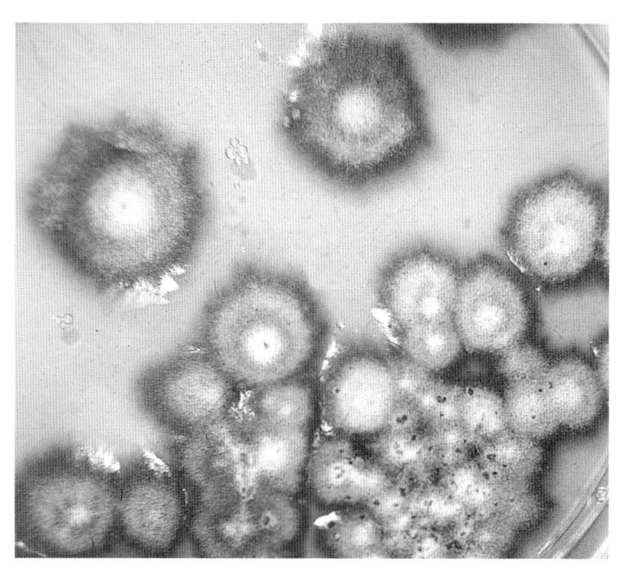

Figure 14.17 Obverse of colonies of *T. tonsurans* appears small, flat, white to brown and suede-like, with feathery edges

Figure 14.18 Obverse of a colony of *T. tonsurans* var. *sulfureum* is yellow, suede-like and crateriform, with a feathery edge

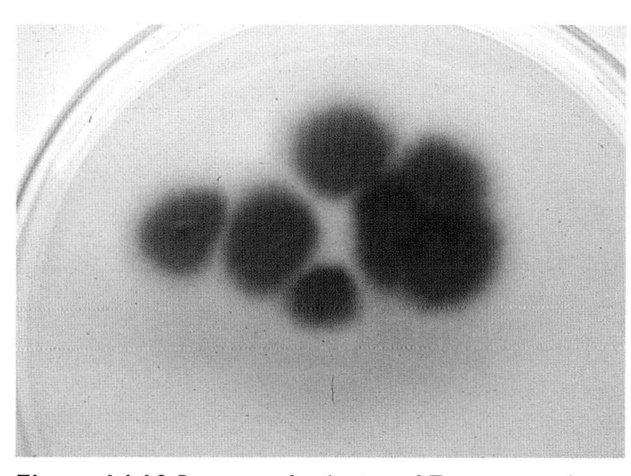

Figure 14.19 Reverse of colonies of *T. tonsurans* shows dark red / brown pigmentation

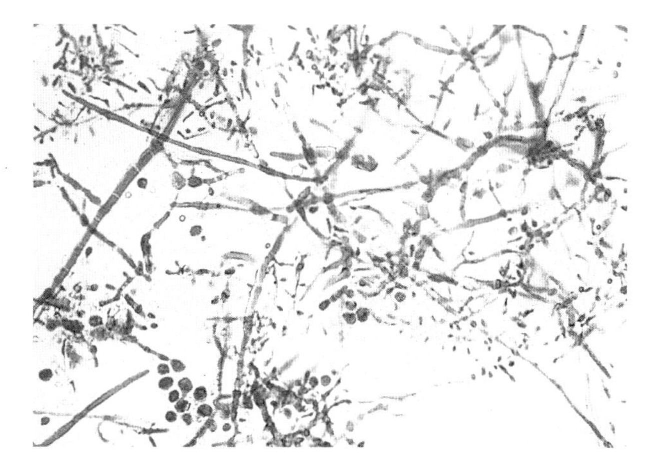

Figure 14.20 Microscopy of *T. tonsurans* shows irregular hyphae; tear-, round, club- and balloon-shaped microconidia; and chlamydoconidia (LPCB)

Epidermophyton floccosum

Source Anthropophilic (worldwide)

Clinical features Infects skin and nail (does not infect hair)

In vivo Hyaline septate hyphae

In vitro

Colony (macroscopically)

Topography: flat with central pucker

Texture: feathery; slightly powdery center

Pigmentation: obverse is yellow-green to dull gray-green (old); reverse is light to medium brown

Colony (microscopically)

Hyphae: regular with numerous chlamydoconidia

Macroconidia: numerous; blunt-ended or club-shaped; smooth-walled

Microconidia: never produced

Special features

None

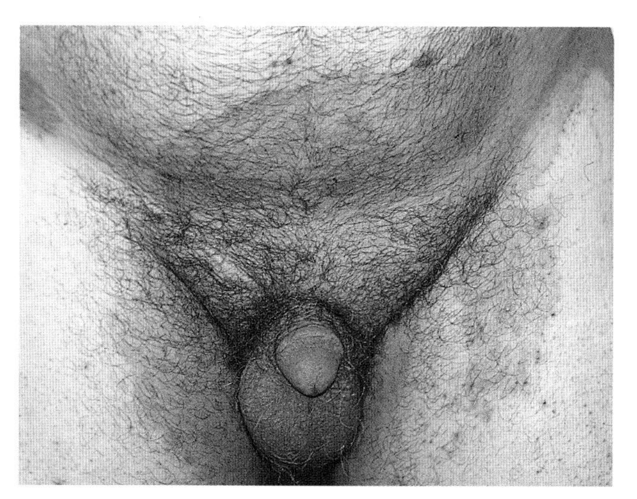

Figure 14.21 Tinea cruris (eczema marginatum) is characterized by well-defined serpiginous, scaly, erythematous lesions

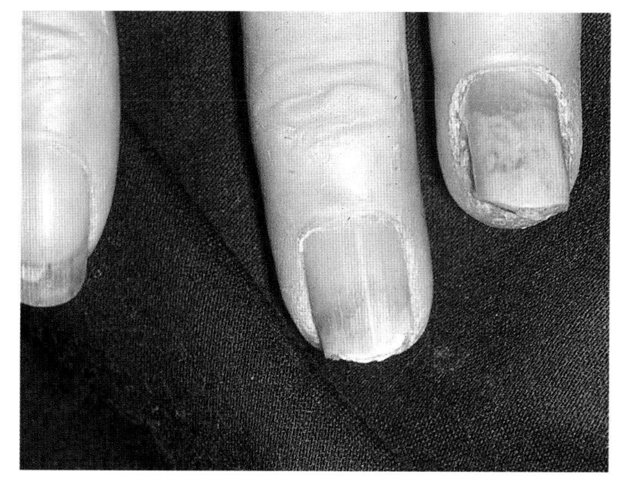

Figure 14.22 Tinea unguium is characterized by distal subungual onychomycosis, with hyperkeratosis, onycholysis and discoloration

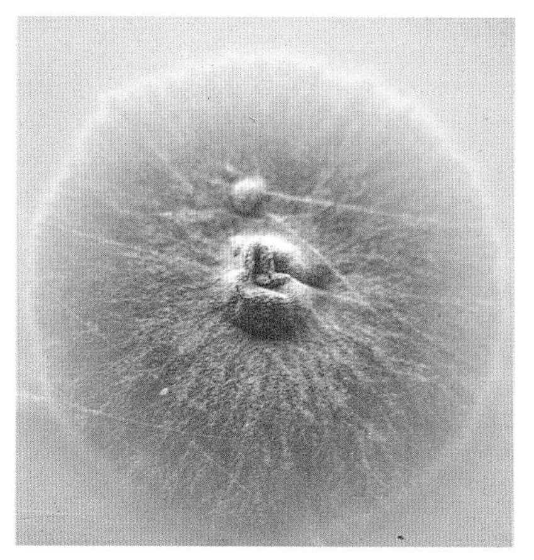

Figure 14.23 Obverse of an *E. floccosum* colony is flat, yellow-green to khaki-gray and feathery, with a puckered center

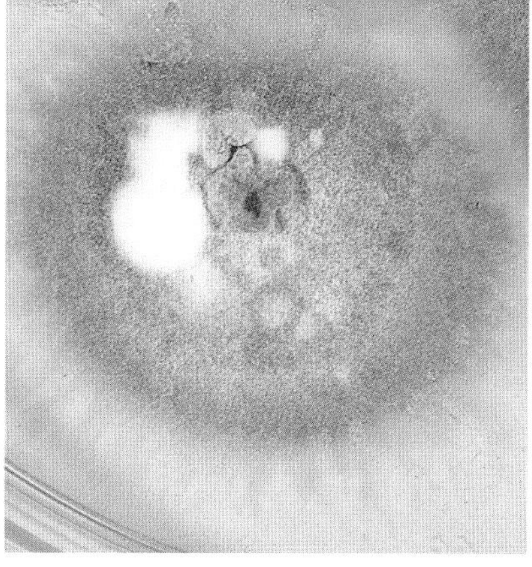

Figure 14.24 Obverse of a polymorphic *E. floccosum* colony shows characteristic flat, white, cottony puffs of sterile mutating mycelia

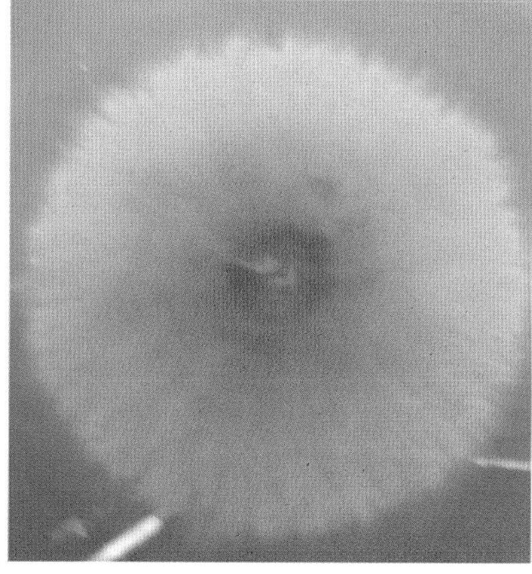

Figure 14.25 Reverse of an *E. floccosum* colony shows light- to medium-brown, non-diffusing pigmentation

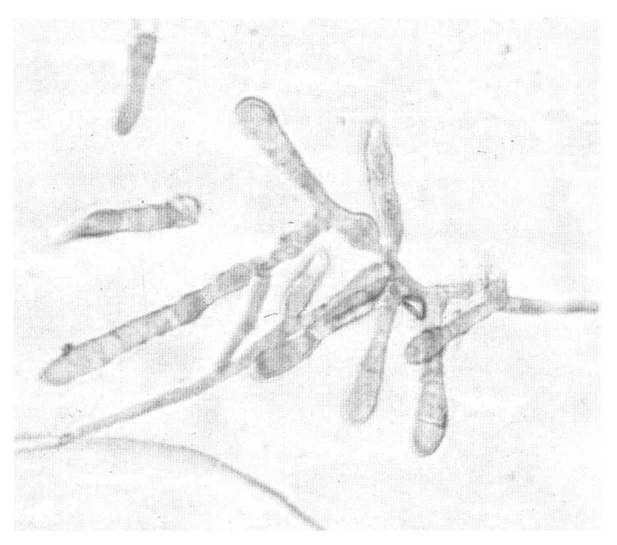

Figure 14.26 Microscopy of *E. floccosum* shows regular hyphae and numerous club-shaped, smooth-walled, 3–5-celled macroconidia

Microsporum canis

Source Zoophilic (in animals worldwide, especially cats and dogs)

Clinical features Infects hair, skin and nails (causes ectothrix fluorescent tinea capitis)

In vivo Skin and nails: septate hyphae; hair: 2–3-μ arthroconidia coating shaft

In vitro

Colony (macroscopically)

Topography: flat, with occasional widespread radial furrows

Texture: coarsely hairy, becoming more dense with age

Pigmentation: obverse is white with rare tan sectors; reverse is non-pigmented to yellow-orange

Colony (microscopically)

Hyphae: regular

Macroconidia: spindle-shaped, thick-walled, echinulate; 6–10 segments

Microconidia: pyriform (non-specific)

Special features

Starch agars: lemon-yellow pigment on PDA

Rice grain test: positive; grows on polished rice

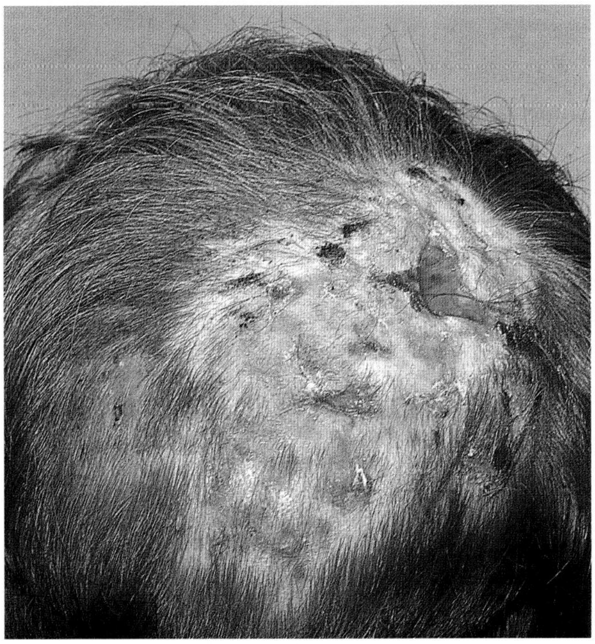

Figure 14.27 Kerion formation with alopecia, erythema, scaling, edema and suppuration caused by *Microsporum canis*

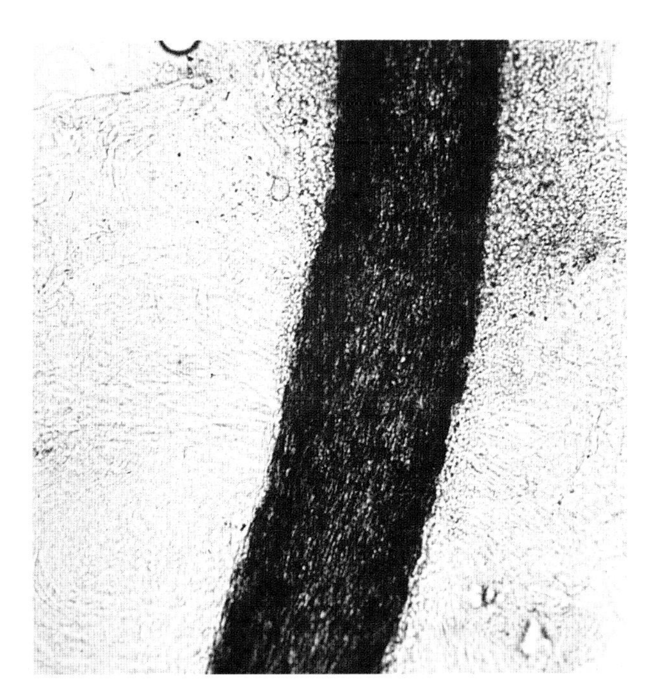

Figure 14.28 Ectothrix tinea capitis. This hair shaft is coated by masses of small (2–3 μ) arthroconidia, seen in a mosaic-tile pattern (KOH)

Figure 14.29 Obverse of an early *M. canis* colony presents as loose, coarsely hairy, white mycelia

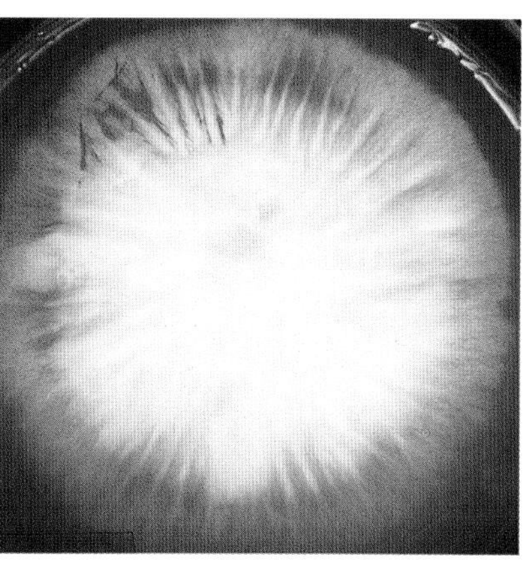

Figure 14.30 Obverse of a mature *M. canis* colony shows a dense, white, hairy to cottony surface with some furrowing

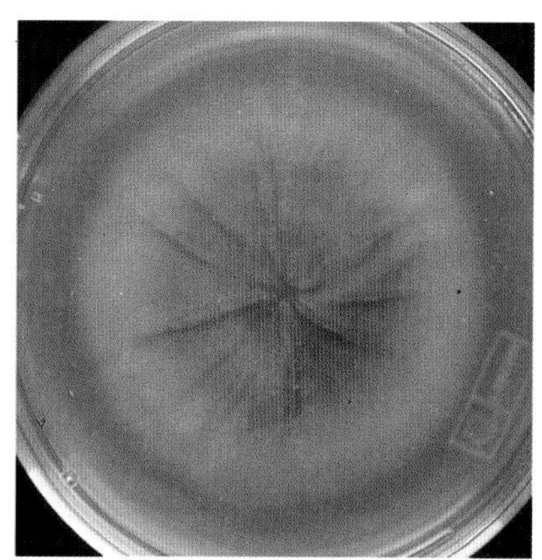

Figure 14.31 Reverse of a *M. canis* colony shows deep-yellow / orange non-diffusing pigmentation

Figure 14.32 Microscopy of *M. canis* shows regular hyphae and pointed, echinulate, thick-walled, 6–10-celled macroconidia (LPCB)

Microsporum gypseum

Source Geophilic (in soil worldwide)

Clinical features Infects hair, skin and nails (skin and non-fluorescent scalp lesions are usually rather inflammatory)

In vivo Skin and nails: septate hyphae; hair: sparse ectothrix

In vitro

Colony (macroscopically)

Topography: usually flat

Texture: coarsely granular (like a chamois cloth)

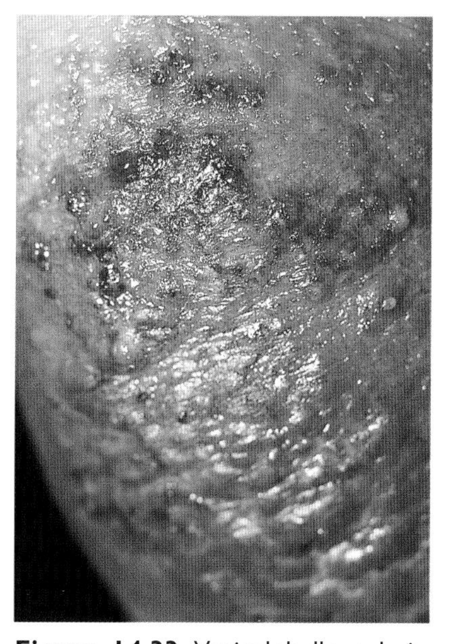

Figure 14.33 Vesiculobullous lesion of inflammatory tinea corporis

Pigmentation: obverse is buff to light-orange; reverse is usually non-pigmented

Colony (microscopically)

Hyphae: regular

Macroconidia: cucumber-shaped, thin-walled, echinulate; 4–6 segments

Microconidia: club-shaped (non-specific)

Special features

None

Figure 14.34 Obverse of a *M. gypseum* colony presents a flat, granular, light-orange, spreading surface

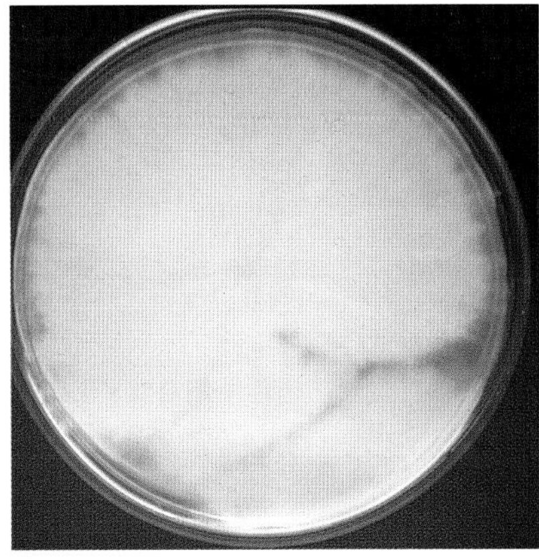

Figure 14.35 Reverse of a *M. gypseum* colony shows no pigmentation

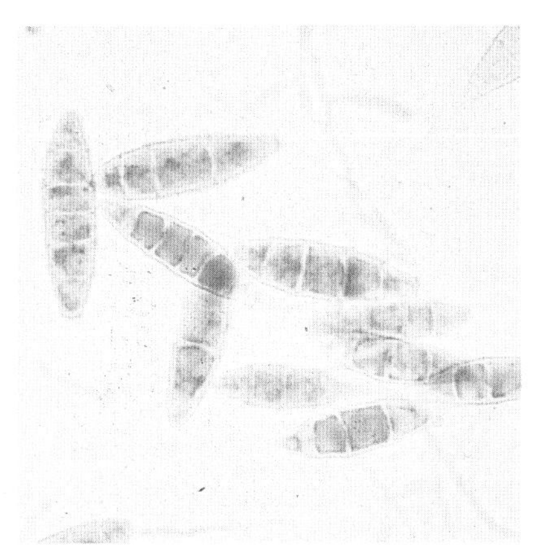

Figure 14.36 Microscopy of *M. gypseum* shows numerous thin-walled, finely echinulate, tapered macroconidia with 3–6 segments (LPCB)

Trichophyton verrucosum

Source Zoophilic (in cattle and, occasionally, sheep and horses worldwide)

Clinical features Infects hair, skin and nails (skin and scalp lesions are usually rather inflammatory)

In vivo Skin and nails: septate hyphae; hair: large ectothrix (non-fluorescent)

In vitro

Colony (macroscopically)

Topography: flat to mounded to folded

Texture: glabrous to slightly downy

Pigmentation: obverse is white to tan (with an occasional yellow strain); reverse is usually non-pigmented

Colony (microscopically)

Hyphae: irregular, with occasional chains of chlamydoconidia

Macroconidia: rare; rat-tail-shaped

Microconidia: rare; club-shaped

Special features

Grows best at 37°C

All strains require thiamine

Some strains require inositol

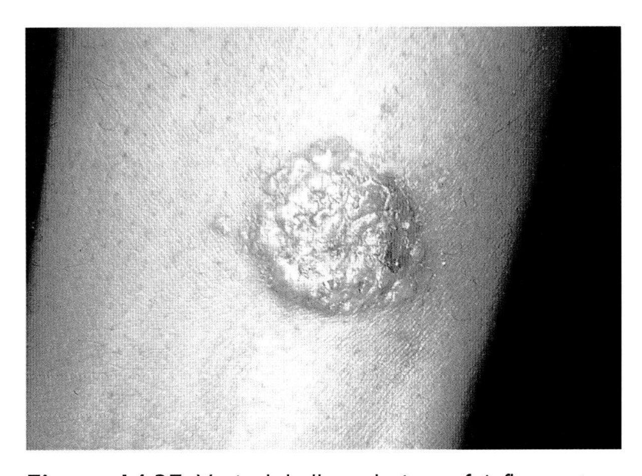

Figure 14.37 Vesiculobullous lesion of inflammatory tinea corporis

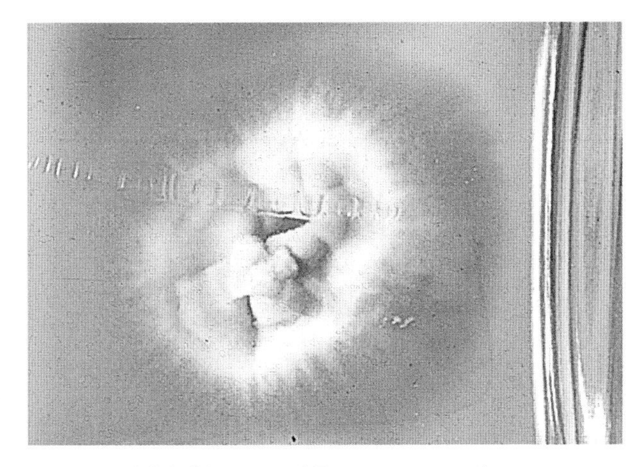

Figure 14.38 Obverse of *T. verrucosum* colony presents a white, glabrous, folded surface

Trichophyton schoenleinii

Source Anthropophilic (Eurasia and Africa; endemic foci in North and South America)

Clinical features Infects hair, skin and nails (favus tinea capitis with scutula and permanent alopecia)

In vivo Skin and nail: septate hyphae; hair: endothrix hyphae (dull-gray fluorescence)

In vitro

Colony (macroscopically)

Topography: folded to cerebriform (often splits agar)

Texture: glabrous to slightly downy

Pigmentation: obverse is white to tan (resembling unbaked pie dough); reverse is usually nonpigmented

Colony (microscopically)

Hyphae: Nail-head hyphae, forming antlers or favic chandeliers

Macroconidia: rarely produced

Microconidia: rarely produced

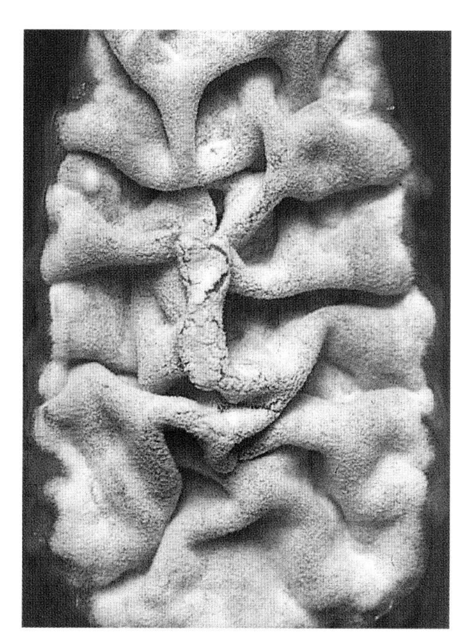

Figure 14.39 Obverse of a *T. schoenleinii* colony presents a white to tan, cerebriform, glabrous surface

Special features

Antler hyphae

Favus with scutula (cup-shaped crusts of debris and hyphae)

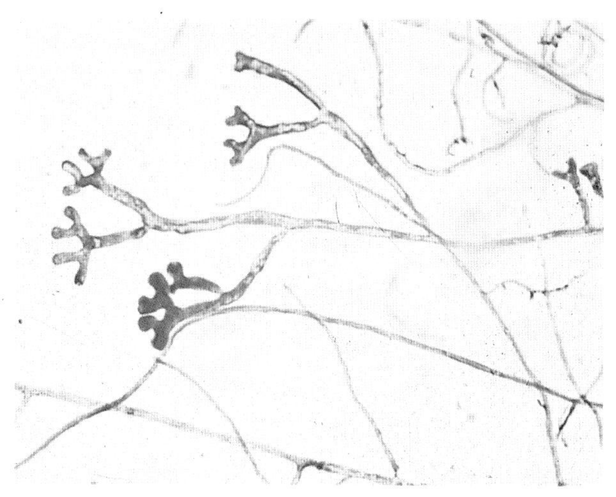

Figure 14.40 Microscopy of *T. schoenleinii* shows irregular hyphae with swollen tips (nail heads) forming antlers or chandeliers

Trichophyton violaceum

Source Anthropophilic (Eastern Europe, Middle East, North Africa; limited endemic areas of Mexico and South America)

Clinical features Infects hair, skin and nails (non-fluorescent, black-dot tinea capitis)

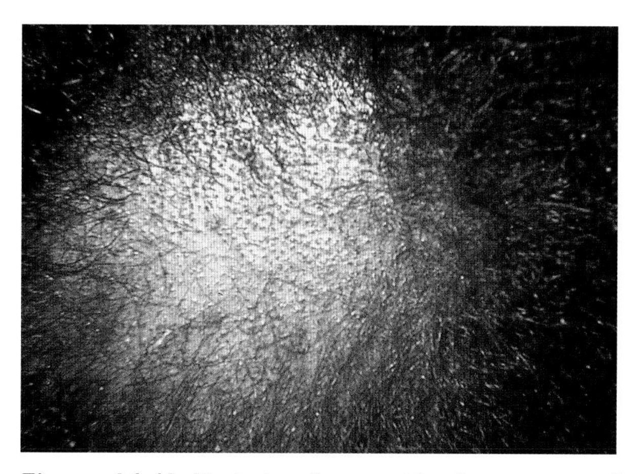

Figure 14.41 Black-dot tinea capitis shows areas of non-inflammatory alopecia with discrete black dots (infected hairs broken off at the follicular orifice)

In vivo Skin and nails: septate hyphae; hair: endothrix arthroconidia (5–8 μ)

In vitro
Colony (macroscopically)
Topography: folded to cerebriform
Texture: glabrous
Pigmentation: obverse and reverse are both dark purple-red

Colony (microscopically)
Hyphae: twisted and of variable diameter
Macroconidia: seldom produced
Microconidia: seldom produced

Special features
Nutritional test: thiamine requirement
Urease test: negative

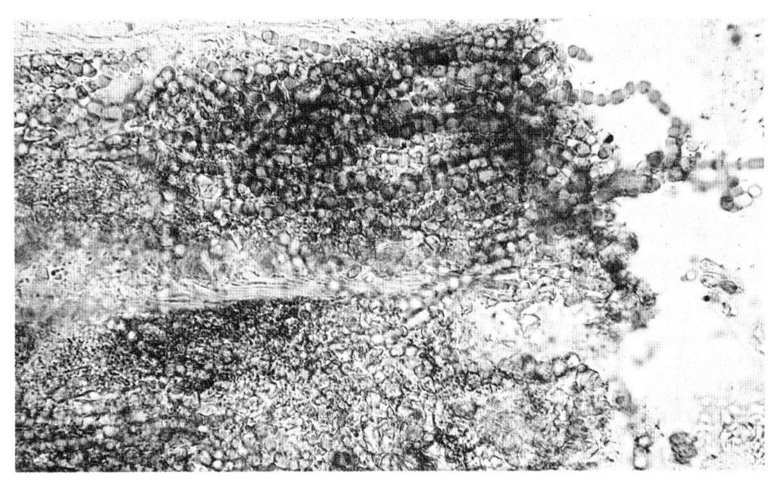

Figure 14.42 Microscopy of endothrix tinea capitis shows an endothrix-infected hair fragment packed with arthroconidia (KOH)

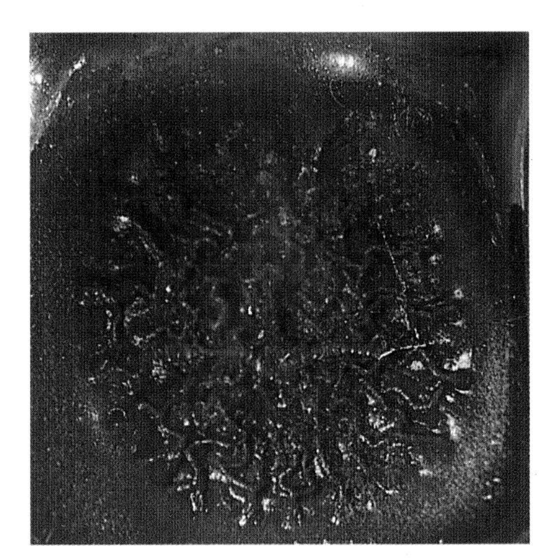

Figure 14.43 Obverse of *T. violaceum* colony is folded, glabrous and violaceous

Trichophyton equinum

Source Zoophilic (species variants found worldwide in horses)

Clinical features Infections of skin and hair have been reported

In vivo Skin: septate hyphae; hair: ectothrix arthroconidia (3–7 μ)

In vitro
Colony (macroscopically)
Topography: flat to mound-like
Texture: cottony
Pigmentation: obverse is white to cream-colored; reverse is dark yellow, becoming reddish-brown; some strains produce a diffusing brown pigment

Colony (microscopically)
Hyphae: regular
Macroconidia: rare; club-shaped, thin-walled, non-echinulate
Microconidia: sparse; pyriform; attached directly to hyphae

Special features
Nutritional test: var. *equinum* requires nicotinic acid for growth
Urease test: negative

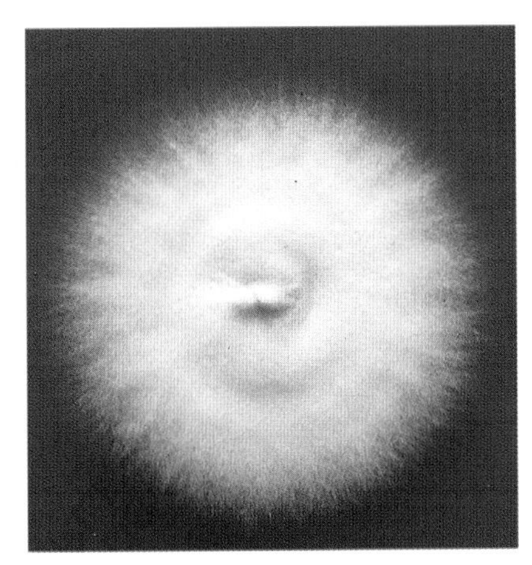

Figure 14.44 Obverse of a *T. equinum* colony, seen on SDA after 14 days, is white, mound-like and cottony

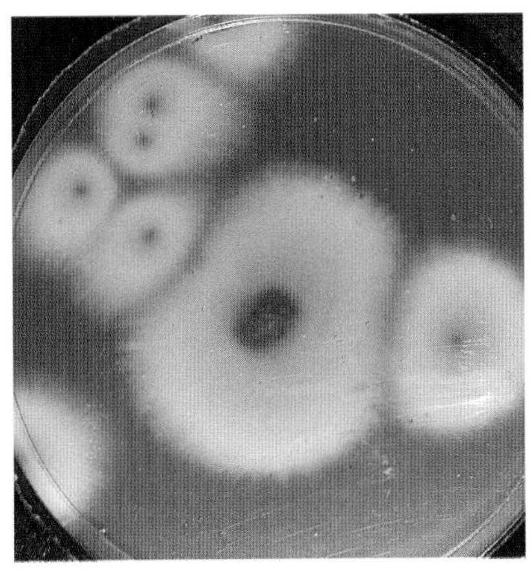

Figure 14.45 Reverse of a *T. equinum* colony is dark yellow, becoming reddish-brown in the center

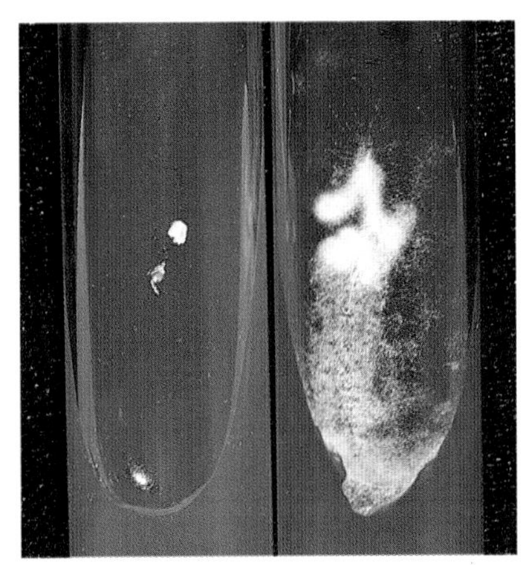

Figure 14.46 Two-tube agar test: *T. equinum* requires nicotinic acid in culture medium (right) for growth

Chapter 15 Non-dermatophytic hyaline and dematiaceous molds

The non-dermatophytic molds comprise a large group of heterogeneous filamentous fungi which, for the most part, are considered saprophytic. Under the right circumstances, many of these species may contribute to human infection. These fungi may be broadly categorized into one of two groups according to their ability to produce melanin. Those organisms incapable of producing this pigment are referred to as 'hyaline' whereas those which produce melanin in their hyphae and / or reproductive structures are called 'dematiaceous'. The latter are sometimes called the 'black fungi'. A number of the more frequently isolated species are described here.

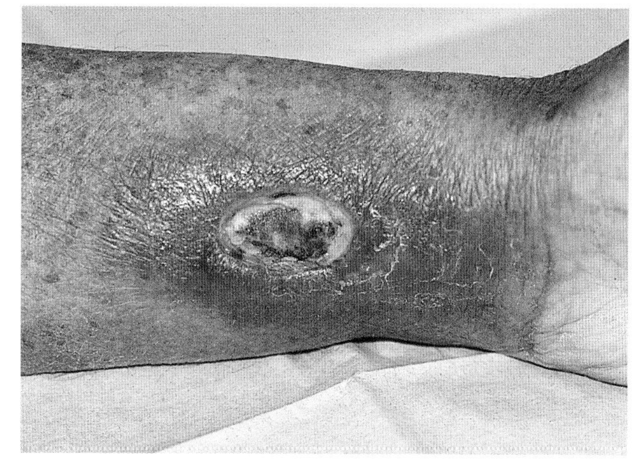

Figure 15.1 Primary cutaneous aspergillosis on the forearm. The necrotic lesion is associated with an intravenous line contaminated with *A. flavus*

HYALINE MOLDS

Aspergillus (A. fumigatus, A. niger, A. flavus, A. terreus)

Source Ubiquitous; isolated frequently from decaying plants and soil

Clinical features Opportunistic pathogens of skin, nails, lung, central nervous system, sinuses, eye, heart, etc.

In vivo Hyaline septate hyphae with bifurcate branching at acute angles (30–60°)

In vitro

Colony (macroscopically)

Topography: flat to folded

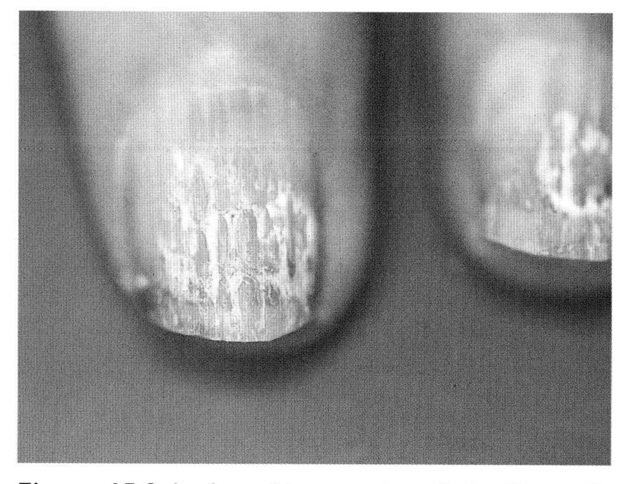

Figure 15.2 Leukonychia mycotica of the fingernails caused by colonization with *A. flavus*

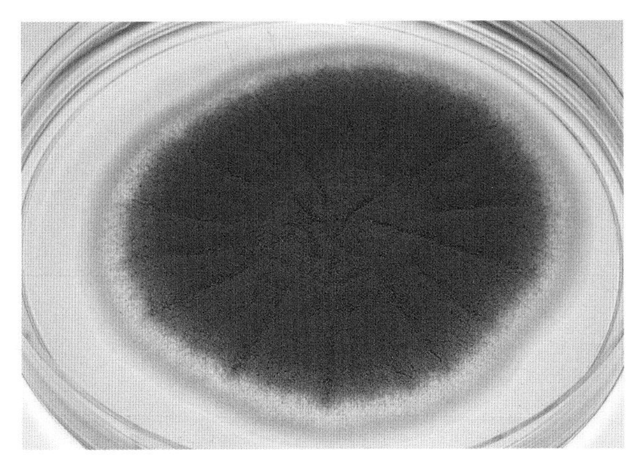

Figure 15.3 Obverse of *A. fumigatus* colony is flat, granular and blue-green after 10 days of growth on SDA

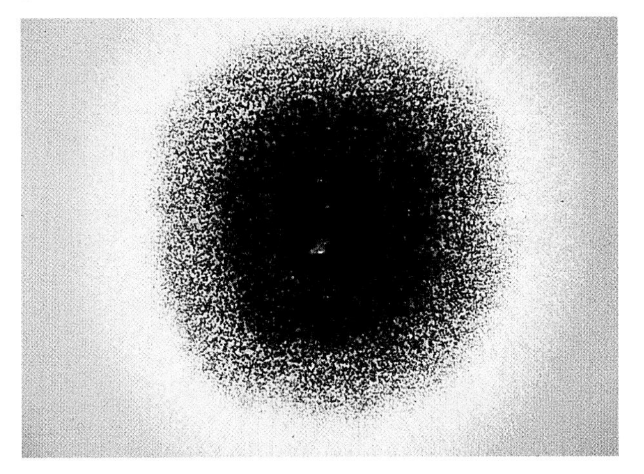

Figure 15.4 Obverse of *A. niger* colony is flat, granular and black after 10 days of growth on SDA

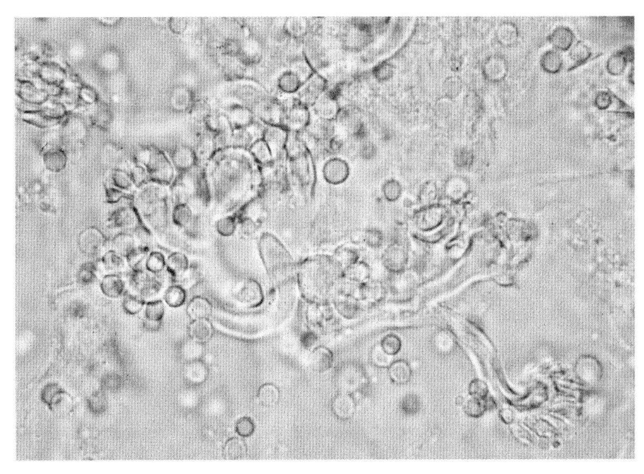

Figure 15.5 Nail scraping in *Aspergillus* onychomycosis shows hyphae, fruiting heads and phialoconidia (KOH)

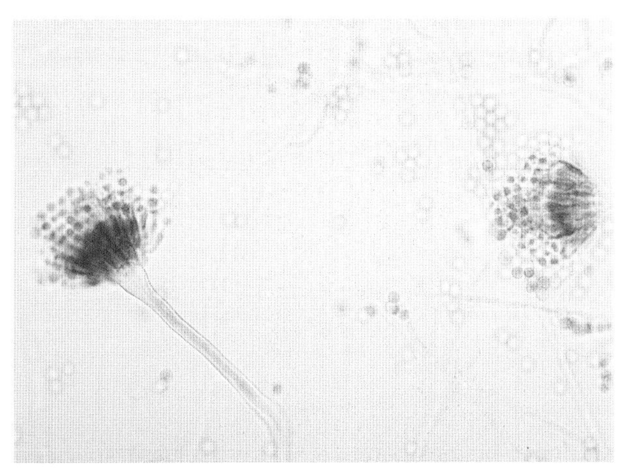

Figure 15.6 Microscopy of *Aspergillus* reveals fruiting heads composed of a conidiophore, vesicle, phialides and phialoconidia (LPCB)

Texture: powdery to granular

Pigmentation: obverse is white, gray, green, blue, gold, black, etc.; reverse is variable; some strains produce diffusing pigment

Colony (microscopically)

Hyphae: regular hyaline septate; form fruiting heads comprising a conidiophore, vesicle and flask-shaped phialides producing phialoconidia

Acremonium (Cephalosporium)

Source Ubiquitous; isolated from decaying plants and soil

Clinical features Reported cause of mycetoma, onychomycosis, ocular mycoses and meningitis

In vivo Hyaline septate hyphae

In vitro

Colony (macroscopically)

Topography: flat to folded

Texture: glabrous to powdery to cottony

Pigmentation: obverse is white, gray, or pink; reverse is colorless to pale yellow

Colony (microscopically)

Hyaline septate mycelia give rise to single, delicate, tapered phialides, which produce oblong phialoconidia that tend to be loosely grouped at its apex

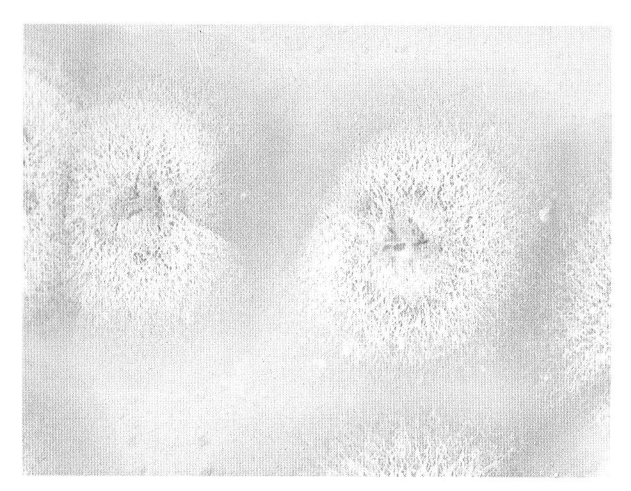

Figure 15.7 Obverse of *Acremonium* colonies are powdery-white to rose-colored after 10 days of growth on SDA

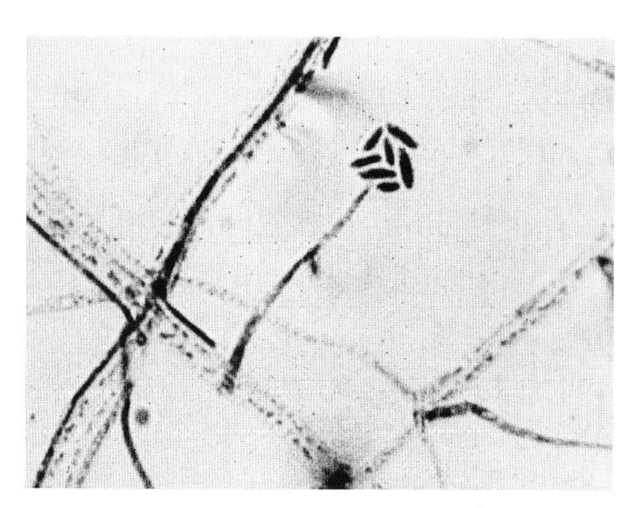

Figure 15.8 Microscopy of *Acremonium* shows hyphae, phialides and clusters of conidia (LPCB)

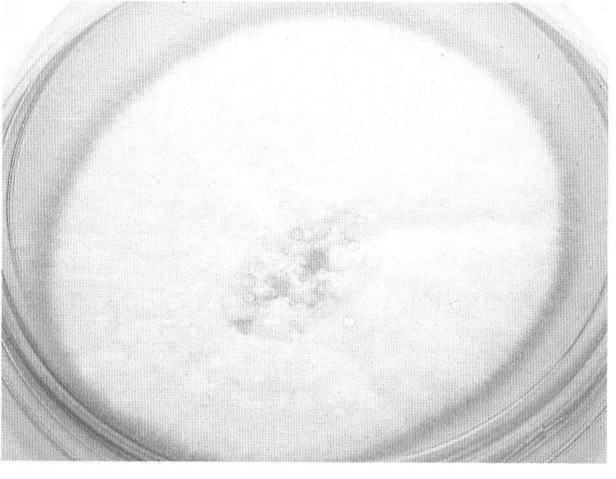

Figure 15.9 Obverse of *Fusarium* colony is wooly, folded and white to brown after 10 days of growth on SDA

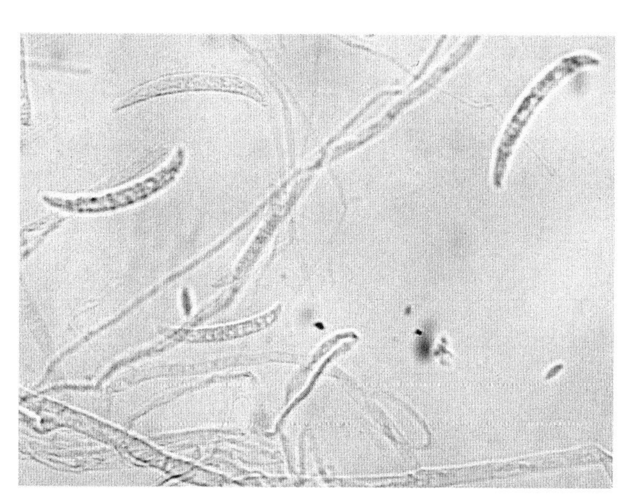

Figure 15.10 Microscopy of *Fusarium* reveals hyphae and canoe-shaped macroconidia

Fusarium

Source Ubiquitous; isolated from decaying plants and soil

Clinical features Reported as a cause of mycetoma, onychomycosis, ocular and pulmonary mycoses

In vivo Hyaline septate hyphae

In vitro
Colony (macroscopically)

Topography: flat to folded

Texture: cottony to wooly

Pigmentation: obverse is white, yellow, red, lavender or brown, depending on strain; reverse is pale yellow, red, lavender or brown

Colony (microscopically)

Hyphae: hyaline and septate

Microconidia: 1–2-celled, ovoid and borne by conidiophores; chlamydoconidia

Macroconidia: multicellular and canoe-shaped; produced by phialides

Geotrichum

Source Ubiquitous; isolated from plants, soil, dairy products; part of normal human flora

Clinical features Possible cause of mucosal thrush-like disease and pulmonary mycosis

In vivo Hyaline septate hyphae

In vitro
Colony (macroscopically)
Topography: flat
Texture: glabrous to powdery to cottony
Pigmentation: obverse is white; reverse is white

Colony (microscopically)
Hyphae: hyaline and septate; disarticulate (non-alternating) into round to rectangular arthroconidia which germinate from one corner

Figure 15.11 Obverse of *Geotrichum* colony is glabrous to powdery and white after 7 days of growth on SDA

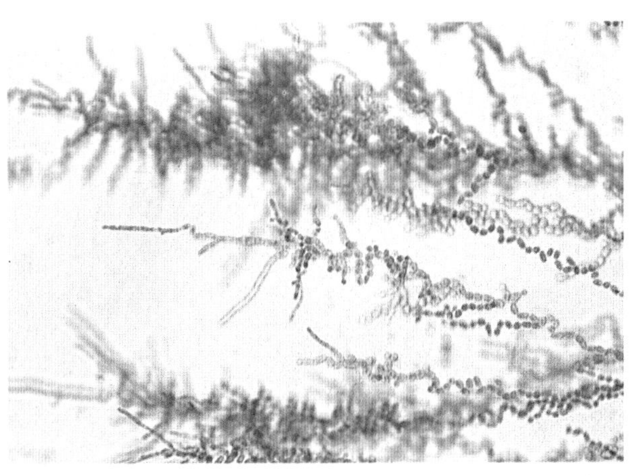

Figure 15.12 Microscopy of *Geotrichum* reveals hyphae separating into arthroconidia (LPCB)

Mucor

Source Ubiquitous; isolated from decaying organic material

Clinical features Less common cause of zygomycosis (rhinocerebral, pulmonary, cutaneous, gastrointestinal and central nervous system)

In vivo Non- to rarely septate, broad, ribbon-like hyphae which branch at 90° angles

In vitro
Colony (macroscopically)
Topography: aerial (fills tube or Petri dish)
Texture: wooly
Pigmentation: obverse is dirty yellow to gray; reverse is white

Colony (microscopically)
Non- to rarely septate hyphae, branched or unbranched sporangiophores with columellae, and sporangia filled with round sporangiospores

Figure 15.13 Obverse of *Mucor* colony is wooly and gray. *Mucor* grows rapidly, as seen here after 5 days on SDA

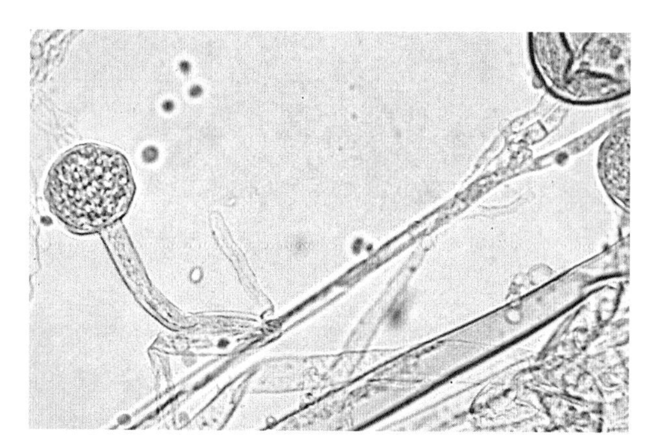

Figure 15.14 Microscopy of *Mucor* shows large hyphae, and sporangiophores bearing sporangia filled with sporangiospores

Figure 15.15 Obverse of *Penicillium* colony is flat, powdery and blue-green after 10 days of growth on SDA

Penicillium

Source Ubiquitous; commonly isolated from many environmental sources

Clinical features Infections of lung, liver, bone, skin and lymphatics (due to *P. marneffei*)

In vivo Hyaline two-celled bodies which multiply by fission

In vitro
Colony (macroscopically)
Topography: flat to plicate
Texture: powdery to granular
Pigmentation: obverse is white, green, blue-green, yellow or gray; reverse is white, tan, red or brown

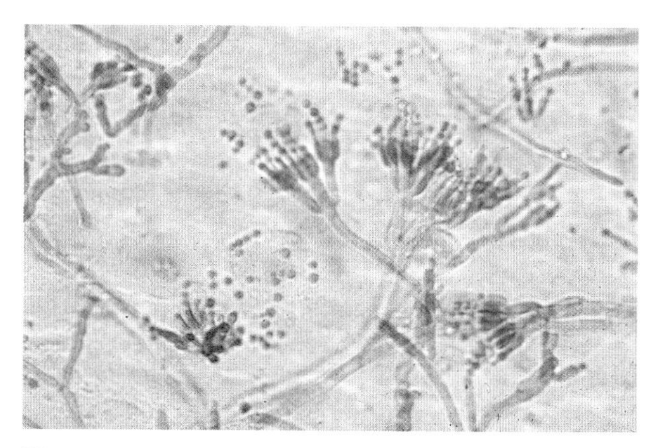

Figure 15.16 Microscopy of *Penicillium* shows hyphae with broom-like fruiting heads (LPCB)

Colony (microscopically)
Hyphae: hyaline and septate
Branched or unbranched conidiophores, with metulae giving rise to whorls of phialides producing chains of phialoconidia

Rhizopus

Source Ubiquitous; isolated from soil and decaying organic material

Clinical features Principal cause of zygomycosis (rhinocerebral, pulmonary, cutaneous, gastrointestinal and central nervous system)

In vivo Non- to rarely septate, broad, ribbon-like hyphae which branch at 90° angles

In vitro
Colony (macroscopically)
Topography: aerial (fills tube or Petri dish)
Texture: wooly
Pigmentation: obverse is white, becoming dirty yellow to gray-brown; reverse is white

Colony (microscopically)
Hyphae: non- to rarely septate
Rhizoids give rise to unbranched sporangiophores, with columellae and sporangia filled with ovoid sporangiospores

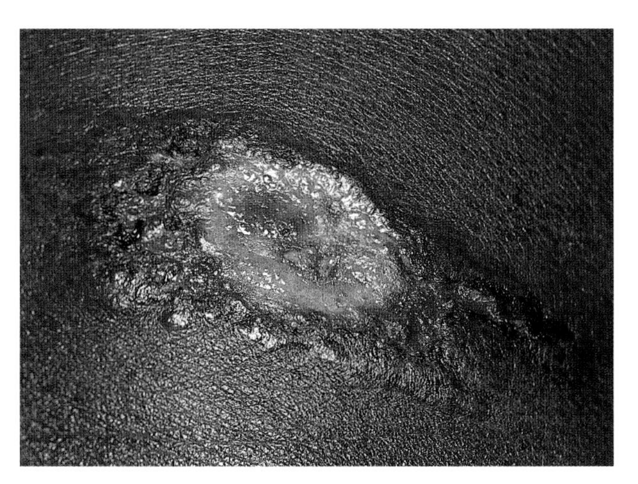

Figure 15.17 Cutaneous zygomycosis presenting as ulceration on the anterior leg due to *Rhizopus* infection

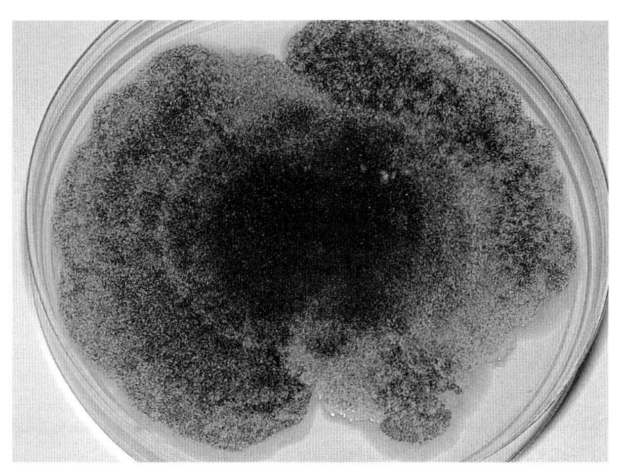

Figure 15.18 Obverse of a *Rhizopus* colony is loose, wooly and gray. *Rhizopus* grows rapidly, as seen here after 5 days on SDA

Scedosporium apiospermum (sexual form: *Pseudallescheria boydii*)

Source Ubiquitous; commonly isolated from soil, manure and decaying plant material

Clinical features Mycetoma and, rarely, ocular, pulmonary, bone and central nervous system infections

In vivo Mycetoma: white grains; other: hyaline septate hyphae

In vitro

Colony (macroscopically)

Topography: flat to mound-like

Texture: cottony to wooly

Pigmentation: obverse is white, becoming gray or dark tan; reverse is brown to black central pigmentation

Colony (microscopically)

Hyphae: hyaline and septate

Unbranched or branched conidiophores, with annellides giving rise to truncate, ovoid annelloconidia

Cleistothecia with ascospores seen in some homothallic strains

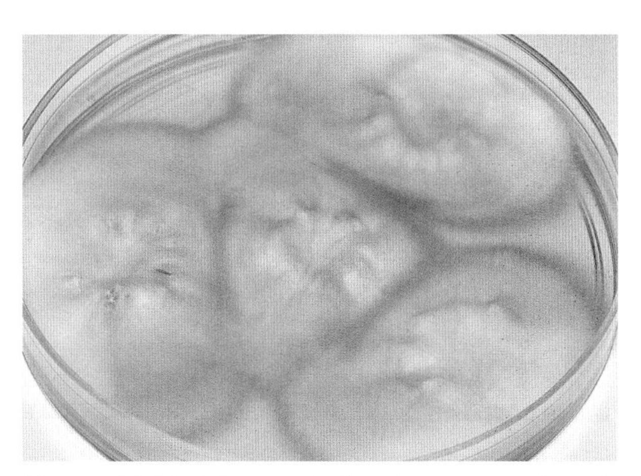

Figure 15.19 Obverse of *S. apiospermum* colony is cottony, tan and mound-like after 10 days of growth on SDA

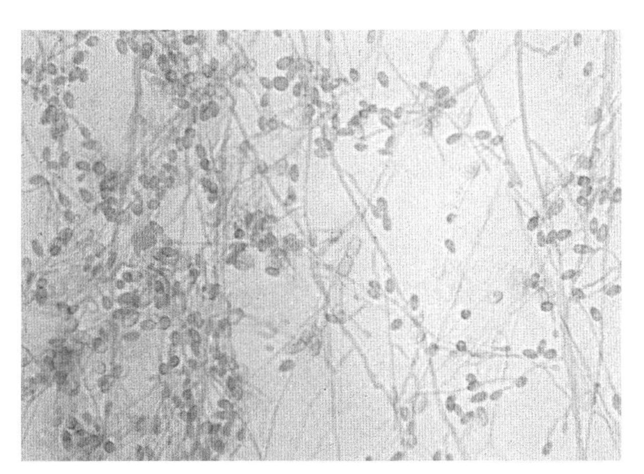

Figure 15.20 Microscopy of *S. apiospermum* showing hyphae and oval annelloconidia (LPCB)

Scopulariopsis

Source Ubiquitous; isolated from soil

Clinical features Non-dermatophytic onychomycosis, rare keratomycosis and pulmonary infections

In vivo Hyaline septate hyphae; occasional annelloconidia (onychomycosis)

In vitro
Colony (macroscopically)
Topography: flat to folded
Texture: granular
Pigmentation: obverse is medium brown, rarely white or black; reverse is tan

Colony (microscopically)
Hyphae: septate
Branched or unbranched conidiophores, with annellides producing chains of large, truncate, rough-walled, lemon-shaped annelloconidia

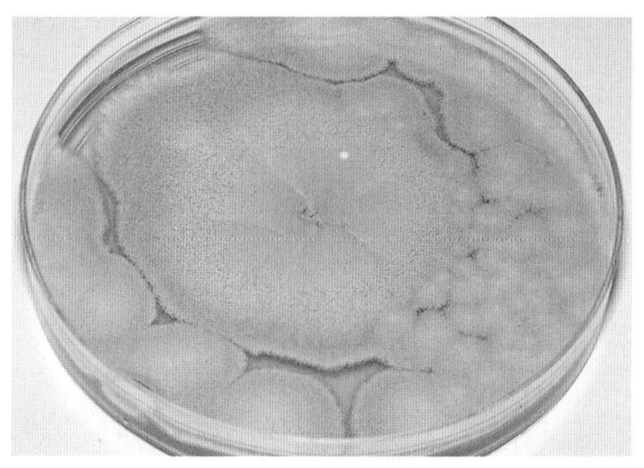

Figure 15.21 Obverse of *Scopulariopsis* colony is flat, granular and brown after 7 days of growth on SDA

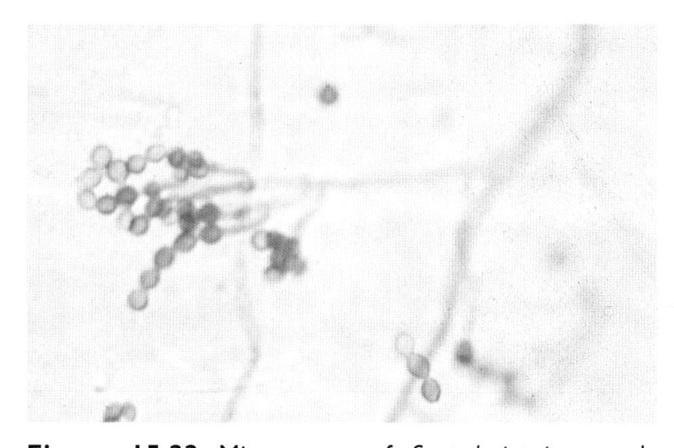

Figure 15.22 Microscopy of *Scopulariopsis* reveals hyphae and oval, lemon-shaped annelloconidia (LPCB)

DEMATIACEOUS MOLDS

Alternaria

Source Ubiquitous; usually isolated from plants and soil

Clinical features Phaeohyphomycosis; rarely, onychomycosis and common mold allergen

In vivo Light to darkly pigmented septate hyphae

In vitro
Colony (macroscopically)
Topography: flat to mound-like or sometimes folded
Texture: cottony to wooly
Pigmentation: obverse is gray to olive-brown to black; reverse is brown to black

Colony (microscopically)
Hyphae: dematiaceous and septate
Conidiophores produce chains of large, dark, muriform (septations in two directions) poroconidia which taper at their distal ends

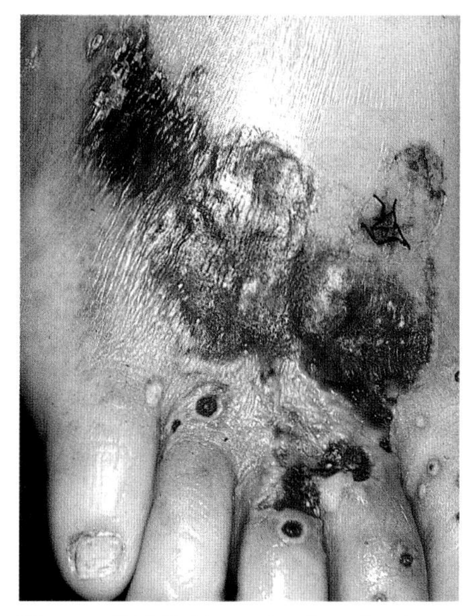

Figure 15.23 *Alternaria* phaeohyphomycosis: small pustules and papules which coalesce and form dark crusts

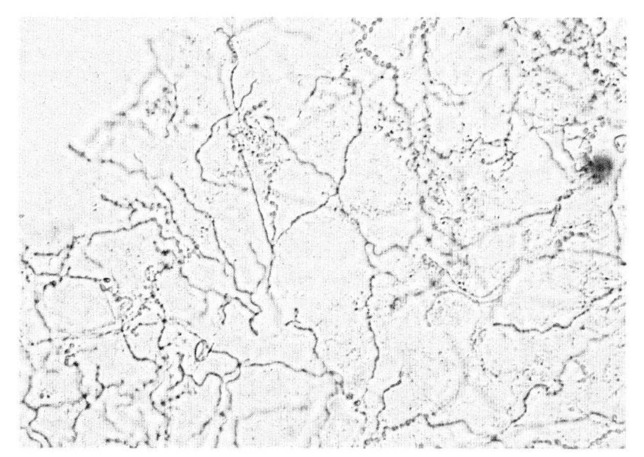

Figure 15.24 Microscopy of a phaeohyphomycotic lesion scraping reveals numerous dark septate hyphae

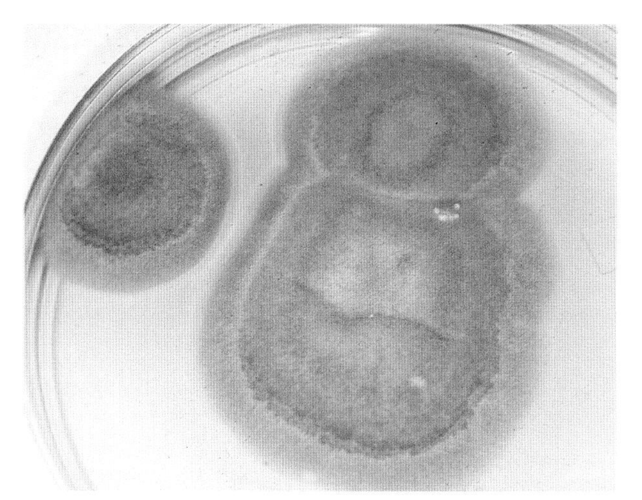

Figure 15.25 Obverse of *Alternaria* colony is dark, cottony and mound-like after 7 days of growth on SDA

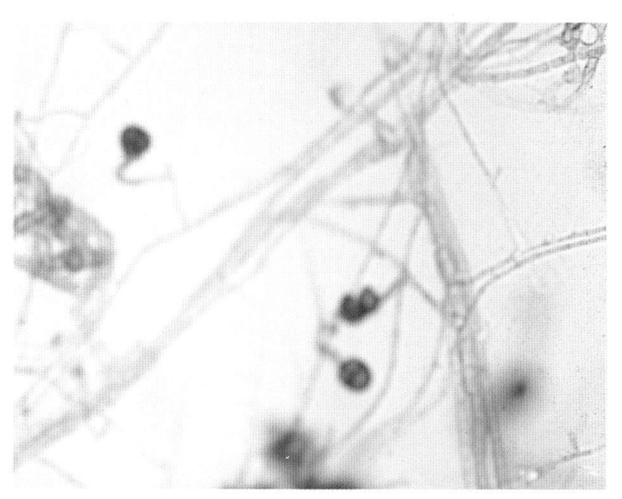

Figure 15.26 Microscopy of *Alternaria* shows dark hyphae and muriform conidia (LPCB)

Bipolaris

Source Ubiquitous; isolated from plants

Clinical features Cutaneous to systemic phaeohyphomycosis involving the skin, eye, bone, heart and central nervous system

In vivo Light to darkly pigmented septate hyphae

In vitro

Colony (macroscopically)

Topography: flat to folded

Texture: cottony

Pigmentation: obverse is olive-brown to black; reverse is black

Colony (microscopically)

Hyphae: dematiaceous and septate

Geniculate conidiophores produce poroconidia of 3–6 segments with a basal hilum (conidia may germinate from the distal segment)

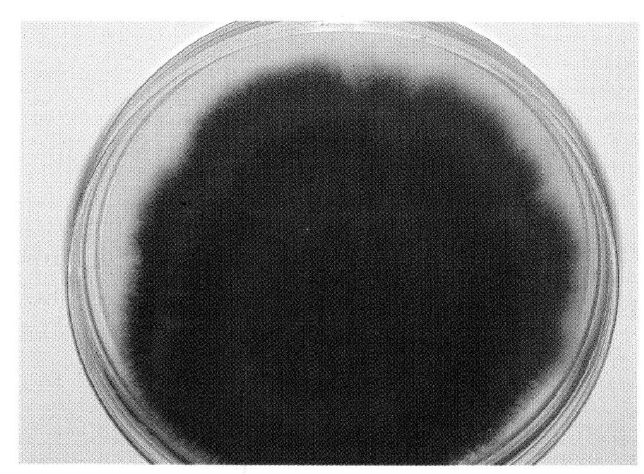

Figure 15.27 Obverse of *Bipolaris* colony is black, cottony and mound-like after 7 days of growth on SDA

Figure 15.28 Microscopy of *Bipolaris* shows dark hyphae and poroconidia

Cladosporium

Source Ubiquitous; usually isolated from plants and soil

Clinical features Chromoblastomycosis (due to *C. carrionii*)

In vivo Sclerotic bodies (medlar bodies, copper pennies)

In vitro

Colony (macroscopically)

Topography: flat or sometimes folded

Texture: velvety

Pigmentation: obverse is olive-brown to black; reverse is brown to black

Colony (microscopically)

Hyphae: dematiaceous and septate

Conidiophores produce delicate branching chains of brown ellipsoidal conidia

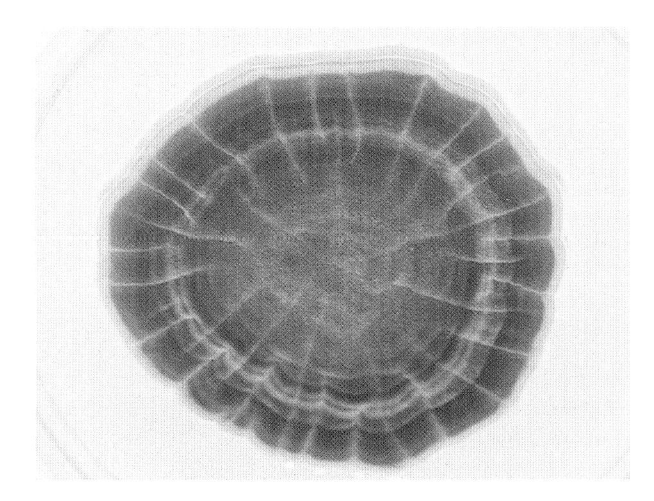

Figure 15.29 Obverse of *Cladosporium* colony is dark, velvety and mound-like after 10 days of growth on SDA

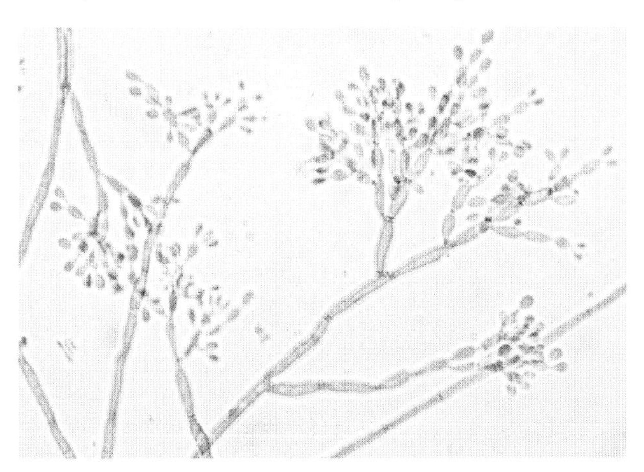

Figure 15.30 Microscopy of *Cladosporium* shows dark hyphae and branching chains of conidia

Curvularia

Source Ubiquitous; isolated from plants and soil

Clinical features Mycetoma and phaeohyphomycosis involving skin, nail, eye, heart, lungs and brain

In vivo Dark grains (mycetoma); light to darkly pigmented hyphae (phaeohyphomycosis)

In vitro

Colony (macroscopically)

Topography: flat to mound-like

Texture: cottony to wooly

Pigmentation: obverse is olive-brown to black; reverse is black

Colony (microscopically)

Hyphae: dematiaceous and septate

Geniculate conidiophores produce poroconidia of 3–4 segments, with a wider, darker, central segment causing the conidia to curve

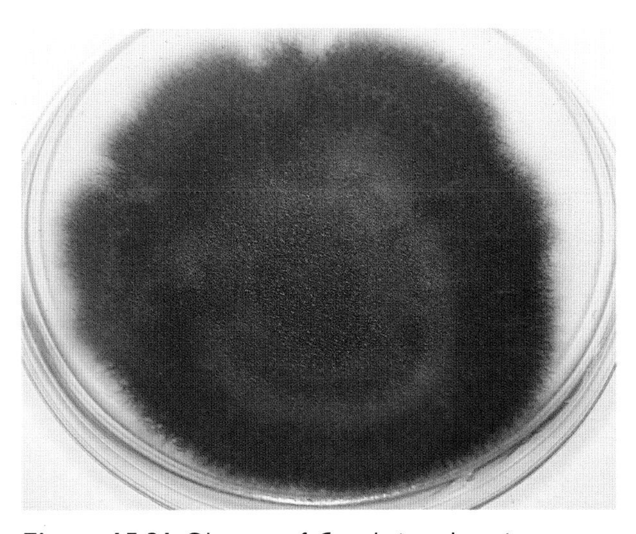

Figure 15.31 Obverse of *Curvularia* colony is cottony, black and mound-like after 7 days of growth on SDA

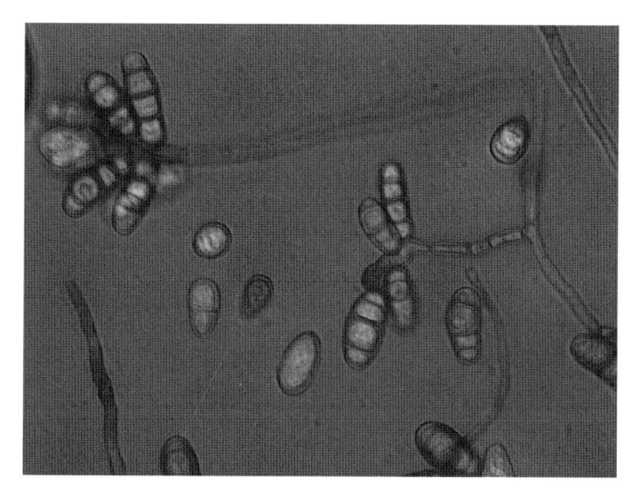

Figure 15.32 Microscopy of *Curvularia* shows dark hyphae and curved poroconidia (LPCB)

Figure 15.33 Obverse of *Fonsecaea* colony is cottony, dark and folded after 10 days of growth on SDA

Fonsecaea

Source Ubiquitous; usually isolated from decaying plants and soil

Clinical features Chromoblastomycosis (*F. pedrosoi, F. compacta*)

In vivo Sclerotic bodies (medlar bodies, copper pennies)

In vitro
Colony (macroscopically)
Topography: flat or sometimes folded
Texture: cottony to wooly
Pigmentation: obverse is olive-brown to brown to gray-black; reverse is brown to black

Colony (microscopically)
Hyphae: dematiaceous and septate
Conidiophores produce sequentially branching chains of brown, round to ellipsoidal conidia (*Cladosporium*-type sporulation). Less commonly, the *Rhinocladiella* or *Phialophora* type of sporulation is also seen

Phaeoannellomyces werneckii (Exophiala werneckii; Cladosporium werneckii)

Source Ubiquitous; isolated from soil and decaying wood

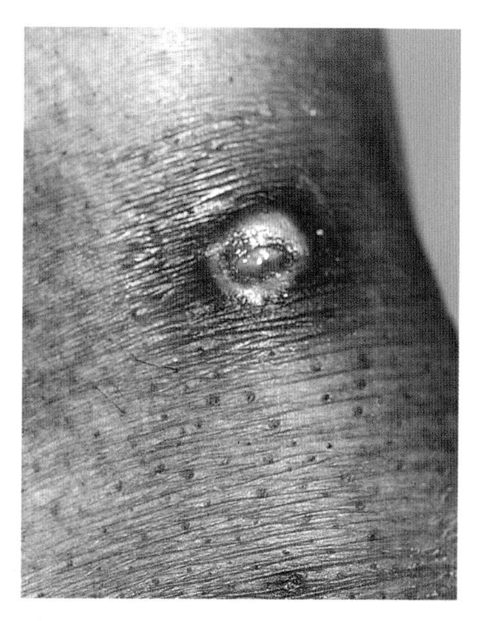

Figure 15.34 Early ulcerated lesion of chromoblastomycosis on the lower leg

Clinical features Tinea nigra (superficial phaeohyphomycosis of palms and soles)

In vivo Dark septate hyphae

In vitro
Colony (macroscopically)
Topography: flat to mound-like
Texture: initially mucoid, becoming velvety
Pigmentation: obverse is dark brown to black; reverse is black

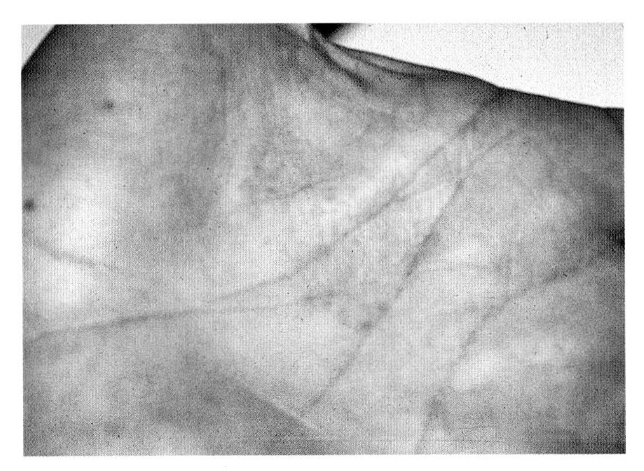

Figure 15.35 Tinea nigra. This lesion on the palm is non-scaly and hyperpigmented

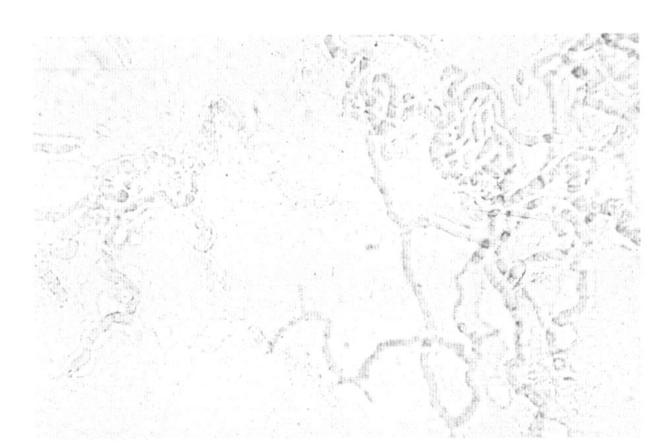

Figure 15.36 Microscopy of a scraping from tinea nigra reveals dematiaceous septate mycelia (KOH)

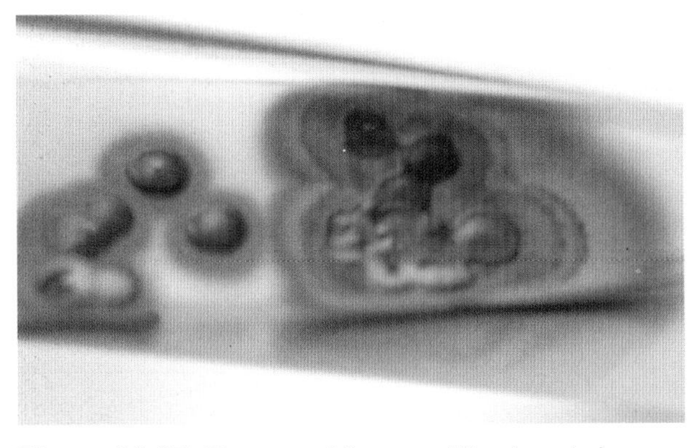

Figure 15.37 Obverse of *P. werneckii* colony is brown-black and cottony after 14 days of growth on SDA

Colony (microscopically)
Dematiaceous bicellular yeast (early) and septate hyphae (later) producing annellides and bicellular annelloconidia

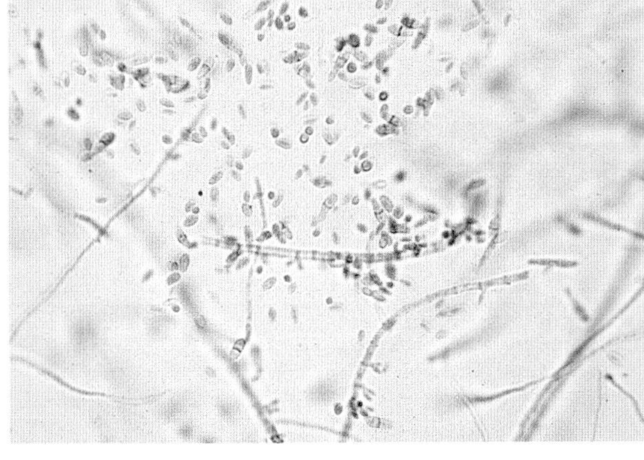

Figure 15.38 Microscopy of *P. werneckii* shows dark hyphae and bicellular annelloconidia (LPCB)

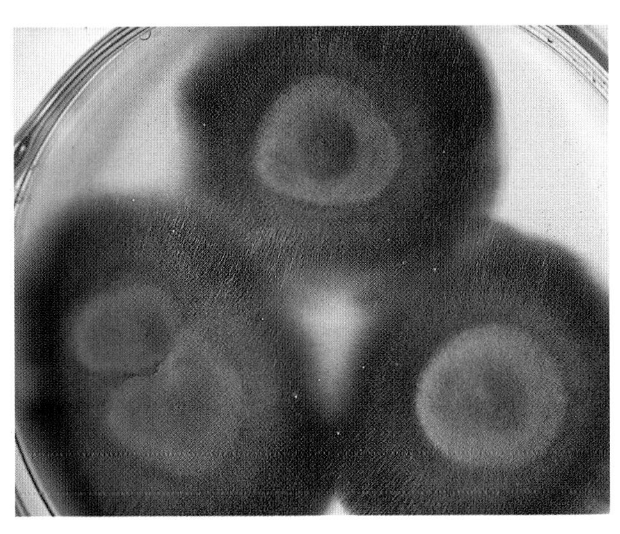

Figure 15.39 Obverse of *Phialophora* colony is cottony, brown and mound-like after 14 days of growth on SDA

Phialophora

Source Ubiquitous; isolated from soil and decaying wood

Clinical features Chromoblastomycosis (*P. verrucosa*), phaeohyphomycosis (*P. richardsiae*)

In vivo Chromoblastomycosis: sclerotic bodies; phaeohyphomycosis: dark septate hyphae

In vitro
Colony (macroscopically)
Topography: flat to mound-like
Texture: cottony to wooly

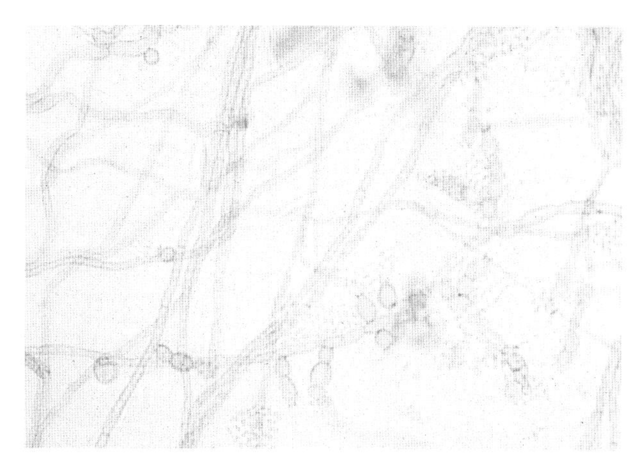

Figure 15.40 Microscopy of *Phialophora* shows dark hyphae, phialides and phialoconidia (LPCB)

Figure 15.41 Obverse of *Rhinocladiella* colony is dark, cottony and mound-like after 14 days of growth on SDA

Pigmentation: obverse is olive-brown to brown to gray-black; reverse is brown to black

Colony (microscopically)

Hyphae: dematiaceous and septate

Phialides: flask-shaped to cylindrical, each bearing a collarette at its opening with an accumulation of phialoconidia ('vase with flowers')

Figure 15.42 Microscopy of *Rhinocladiella* shows dark hyphae and knotted-club conidiation (LPCB)

Rhinocladiella aquaspersa

Source Ubiquitous; usually isolated from decaying plants and soil

Clinical features Chromoblastomycosis

In vivo Sclerotic bodies (medlar bodies, copper pennies)

In vitro

Colony (macroscopically)

Topography: flat or sometimes folded

Texture: cottony to wooly

Pigmentation: obverse is olive-brown to black; reverse is black

Colony (microscopically)

Hyphae: dematiaceous and septate

Conidiophores: cylindrical and unbranched, producing clavate conidia connected by denticles to the tip and sides of the distal stalk (knotted club)

Chapter 16 Yeasts

Fungi which demonstrate a predominantly or exclusively yeast morphology are characteristically round to oval and usually single-celled, and reproduce through budding or fission. Colonies may be smooth, pasty, creamy, mucoid or membranous and range in color from white to cream to pink to red to black.

Species identification is based upon these morphological features as well as the ability to demonstrate pseudohyphae, germ tubes, hyphae, capsules and chlamydospores. Carbohydrate assimilation and fermentation, urease activity, caffeic acid response, and the ability to grow in the presence of cycloheximide and at various temperatures constitute physiological test mechanisms which may confirm the identity of the yeast. Direct microscopy of a patient's specimen may provide the clinician with more immediate information as to the cause of the yeast infection, as suggested by the size, shape, budding or presence of a capsule or pseudohyphae.

Candida albicans (occasional pathogens: *C. tropicalis; C. parapsilosis; C. lusitaniae; C. krusei; C. guilliermondii; C. glabrata;* and *C. pseudotropicalis*)

Source Ubiquitous: any organic substrate; endogenous: colonize mucosal surfaces

Clinical features Mucosal, epidermal, subcutaneous and systemic mycoses

In vivo Hyaline blastoconidia, pseudohyphae and true hyphae

In vitro

Colony (macroscopically)

Topography: flat to mound-like
Texture: smooth or pasty
Pigmentation: obverse is white; reverse is white

Colony (microscopically)

Single, non-encapsulated, budding yeast (blastoconidia); pseudohyphae

Special features

CMA + Tween 80: produces chlamydospores (*C. albicans*)

Germ tube test: positive formation of germ tubes (*C. albicans*)

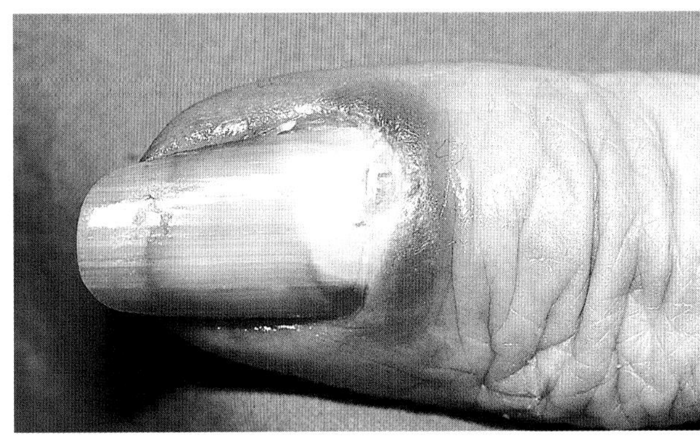

Figure 16.1 Onychomycosis due to *C. albicans*. There is periungual erythema and white discoloration of the nail plate

Sucrose assimilation: positive color change from purple to yellow (*C. albicans*)

Urease test: no urease activity at 4 days

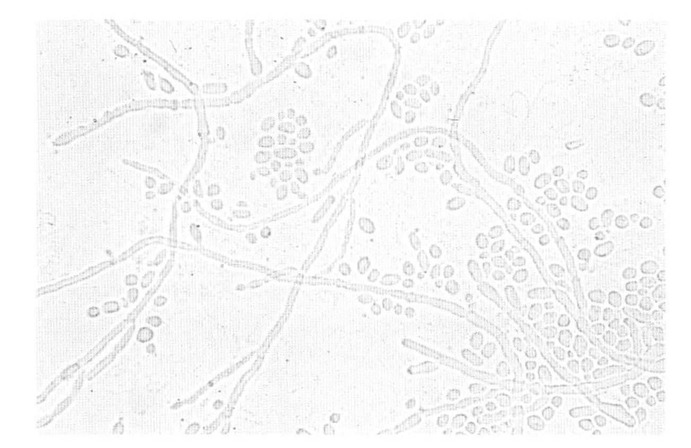

Figure 16.2 Microscopy of candidiasis lesion exudate shows pseudohyphae and blastoconidia (KOH)

Figure 16.3 Obverse of *C. albicans* colony is white, glabrous and pasty after 4 days of growth on SDA

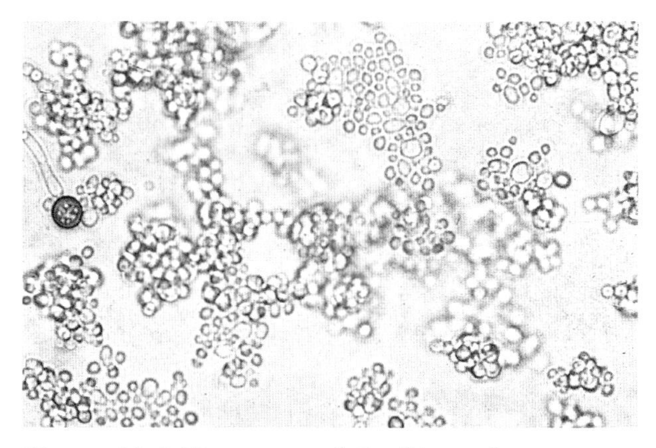

Figure 16.4 Microscopy of *C. albicans* shows masses of blastoconidia and a pseudohypha (LPCB)

Cryptococcus neoformans

Source Ubiquitous; plants, fruit, soil, pigeon feces

Clinical features Mycoses of the lung, skin and central nervous system

In vivo Hyaline encapsulated yeast (5–12 μ)

In vitro

Colony (macroscopically)

Topography: flat to mound-like

Texture: mucoid

Pigmentation: obverse is tan to yellow; reverse is tan to yellow

Colony (microscopically)

Single, encapsulated, budding yeast

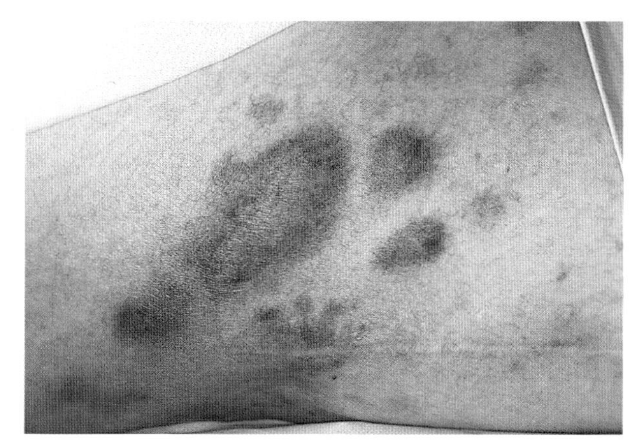

Figure 16.5 Disseminated cutaneous cryptococcosis presents as violaceous nodules on the skin

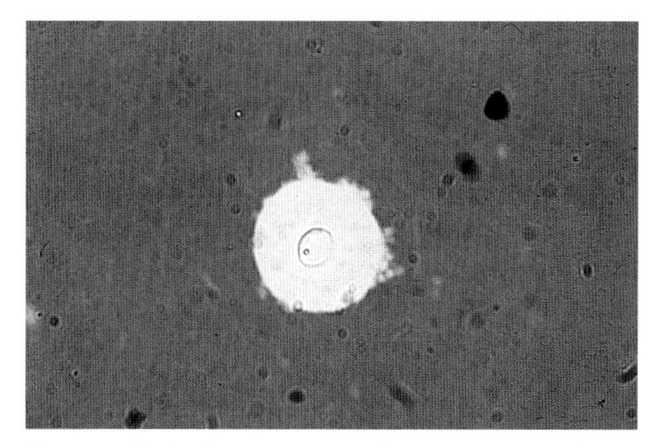

Figure 16.6 Cryptococcosis lesion exudate shows encapsulated yeast (India ink)

Special features

Caffeic acid: brown colony is produced on birdseed agar

Growth at 37°C: positive

Urease test: urease activity within 4 days

Cyclohexamide: inhibits growth

Pseudohyphae: rarely produced

Malassezia furfur (Pityrosporum orbiculare; P. ovale)

Source Endogenous; normal skin flora in 92% of humans

Clinical features Pityriasis versicolor, *Malassezia* folliculitis, seborrheic dermatitis, fungemia

In vivo Hyaline non-encapsulated yeast (3–8 μ); short septate hyphae

In vitro

Colony (macroscopically)

Topography: flat to mound-like

Texture: glabrous

Pigmentation: obverse is white; reverse is white

Colony (microscopically)

Single, non-encapsulated, oval or bottle-shaped, budding yeast

Special features

Oleic acid: requires olive oil on agar surface for growth

Growth at 37°C: grows more readily at body temperature

Cyclohexamide: is able to grow in the presence of this inhibitor

Pseudohyphae: only produces true hyphae

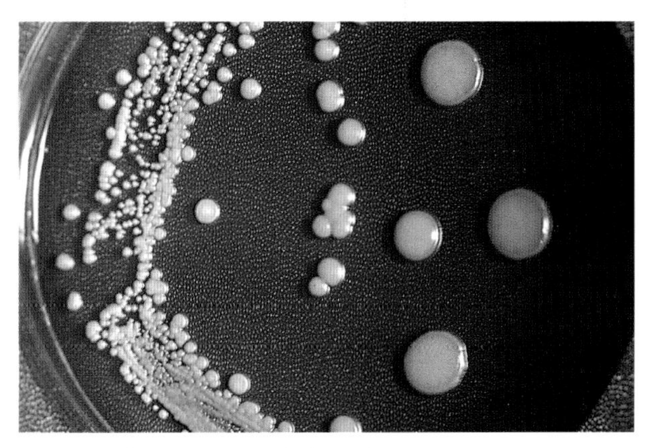

Figure 16.7 Obverse of *C. neoformans* colony is tan and mucoid (slimy) after 5 days of growth on SDA

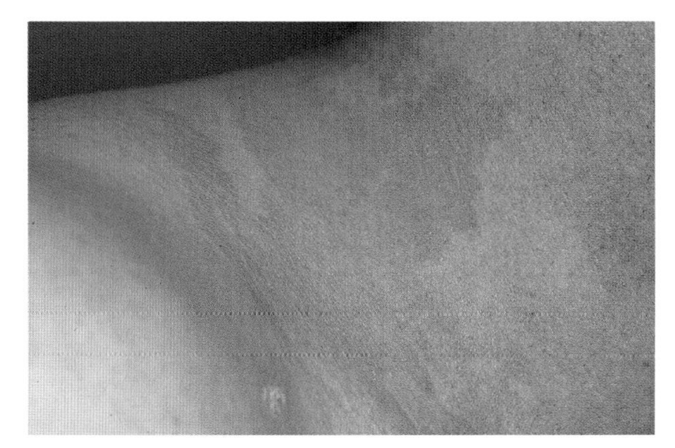

Figure 16.9 Pityriasis versicolor presenting as hyperpigmented, minimally scaling, lesions on the neck

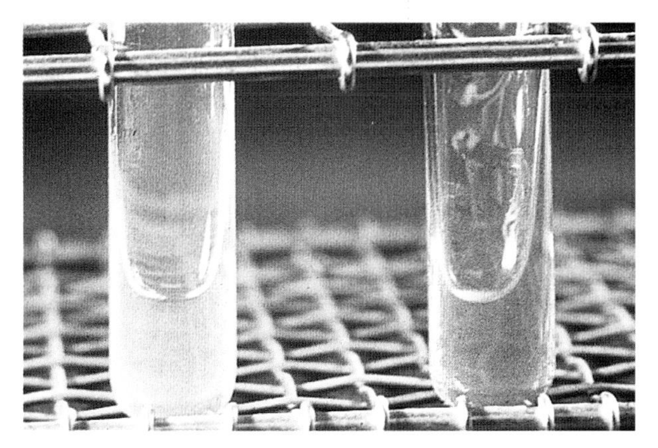

Figure 16.8 Urease test: Positive conversion of Christensen's agar to pink / red is typical of *C. neoformans*

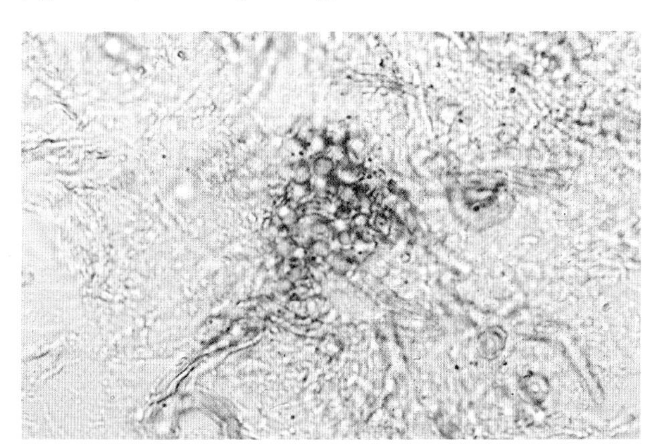

Figure 16.10 Microscopy of pityriasis versicolor lesion shows hyphae and yeast ('spaghetti and meatballs'; KOH)

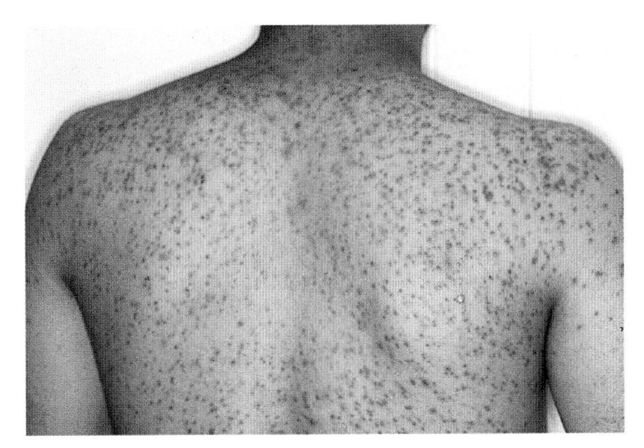

Figure 16.11 *Malassezia* folliculitis of the upper body caused by colonization by *M. furfur*

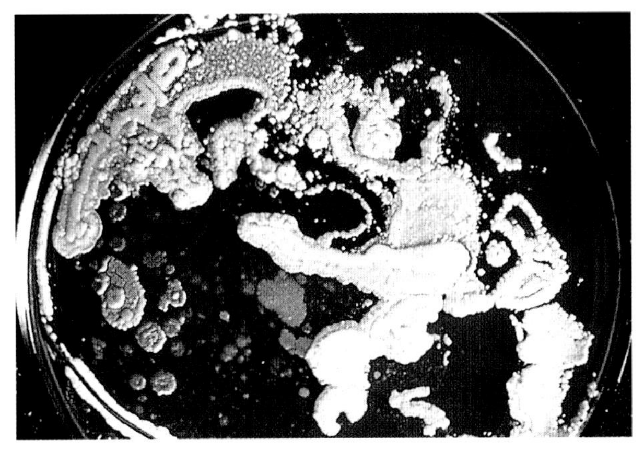

Figure 16.13 Obverse of *M. furfur* colony is white and glabrous after 7 days of growth on SDA with oil at 37°C

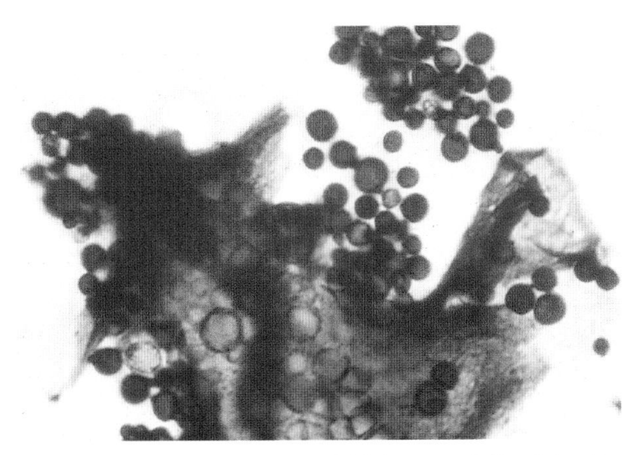

Figure 16.12 Microscopy of *Malassezia* folliculitis skin scraping reveals large clusters of gram-positive yeast

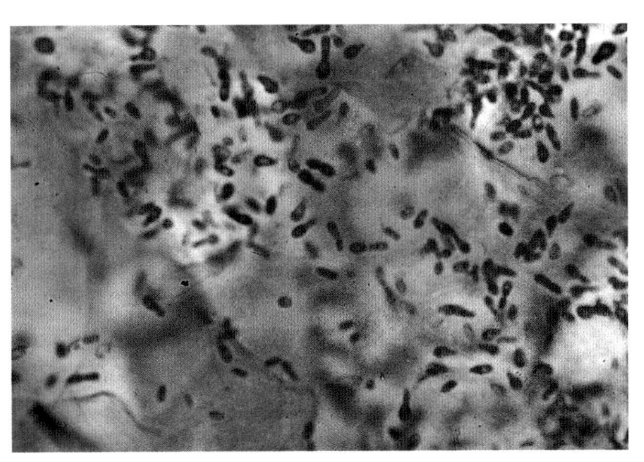

Figure 16.14 Culture microscopy of *M. furfur* shows oval and bottle-shaped blastoconidia (LPCB)

Rhodotorula

Source Ubiquitous airborne contaminant; in soil, water, dairy products

Clinical features Rare pulmonary and systemic mycoses (fungemia, endocarditis, meningitis)

In vivo Hyaline, ovoid to elongate, encapsulated yeast ($2–5 \times 2–14\ \mu$)

In vitro

Colony (macroscopically)

Topography: flat to mound-like

Texture: mucoid

Pigmentation: obverse is pink to coral-red; reverse is pink to coral-red

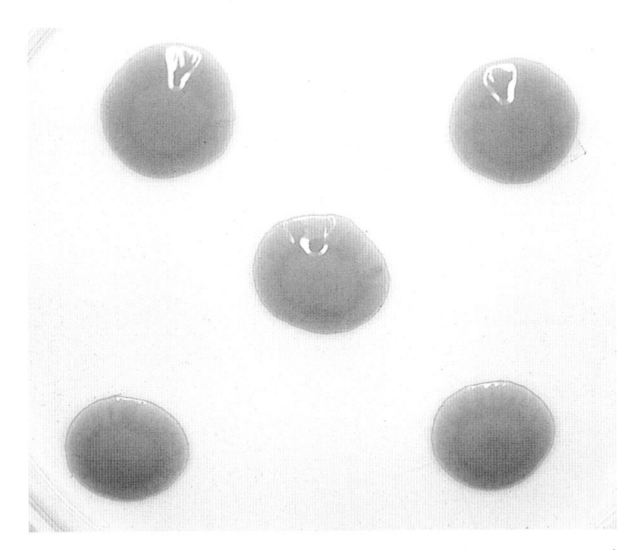

Figure 16.15 Obverse of *Rhodotorula* colony is mucoid and coral-red after 4 days of growth on SDA

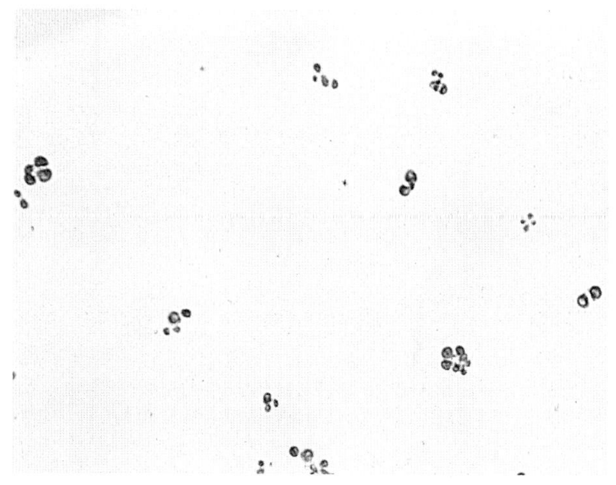

Figure 16.16 Colony microscopy of *Rhodotorula* shows round to oval blastoconidia (LPCB)

Colony (microscopically)

Single to chained, encapsulated, oval, budding yeast

Special features

Urease test: demonstrates urease activity
Growth at 37°C: usually thermotolerant
Cyclohexamide: growth is usually inhibited
Pseudohyphae: very rarely produced

Trichosporon beigelii (T. cutaneum)

Source Ubiquitous; soil reservoir

Clinical features White piedra, onychomycosis, rare systemic mycoses

In vivo Hyaline septate hyphae and budding yeast

In vitro

Colony (macroscopically)

Topography: wrinkled; flat to heaped
Texture: glabrous to powdery
Pigmentation: obverse is cream to gray; reverse is cream to gray

Colony (microscopically)

Hyaline septate hyphae with lateral blastoconidia (yeast)

Hyphae separate into arthroconidia in older colonies

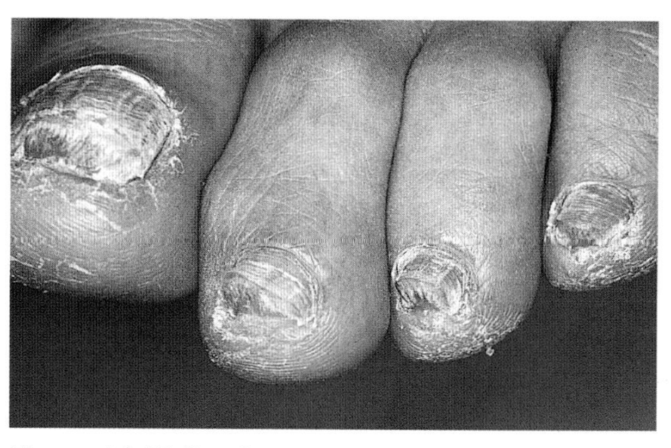

Figure 16.17 Onychomycosis due to *T. beigelii*: Infected nails are hyperkeratotic, onycholytic and discolored

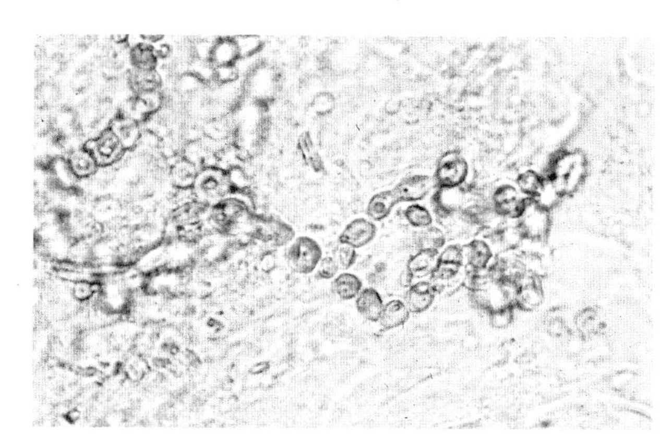

Figure 16.18 Microscopy of *T. beigelii* onychomycosis reveals irregular septate hyphae (KOH)

Figure 16.19 Obverse of *T. beigelii* colony is glabrous, folded and cream-colored after 10 days of growth on SDA

Special features

Urease test: shows positive urease activity
Growth at 37°C: usually thermotolerant
Cyclohexamide: growth not inhibited
Pseudohyphae: produced on CMA

Chapter 17 Dimorphic fungal pathogens

Infectious fungi which demonstrate more than one morphology are referred to as 'dimorphic fungal pathogens'. Organisms which demonstrate one morphology in tissue and another distinct morphology *in vitro* are called 'tissue dimorphic'. An example of such a pathogen is *Coccidioides immitis*, which is found in the form of a mold in nature or on a culture plate, but presents as a spherule in human tissue.

Fungi which grow *in vitro* in a mold form at 25°C, but in a yeast form at 37°C, are called 'thermodimorphic'. Examples include *Sporothrix schenckii*, *Blastomyces dermatitidis*, *Paracoccidioides brasiliensis* and *Histoplasma capsulatum*. Thermodimorphic fungi also demonstrate their yeast morphology in human tissue.

Except for *S. schenckii*, all of the above-mentioned dimorphic fungi are highly infectious and usually use the lungs as the portal of entry. Isolation and identification of these pathogen colonies is dangerous and should only be attempted in a biological safety cabinet by trained personnel. The clinician may safely examine clinical specimens *via* direct microscopy of saline suspensions, KOH preparations, stained smears and touch imprints. Histopathological support of suspected clinical diagnoses may also be of value.

Blastomyces dermatitidis

Source Endemic; probably a soil reservoir and proliferates only under special environmental conditions; seen primarily in North America from the Mississippi River valley to the eastern seaboard

Clinical features Blastomycosis: lung, bone, genitourinary tract; skin lesions are common and present as verrucous plaques with a raised 'stadium' edge

In vivo Hyaline, non-encapsulated, thick (double-contour)-walled yeast (8–20 μ) which undergoes single budding from a broad base (described as resembling a figure-of-eight)

In vitro

Colony (macroscopically)

At 25°C: mold phase (white to tan, cottony to granular)

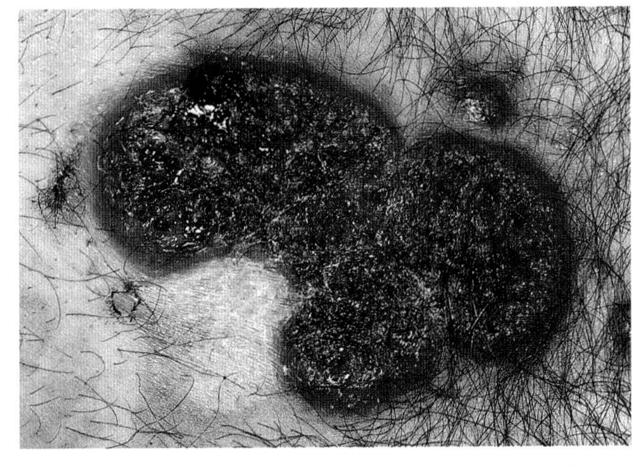

Figure 17.1 Blastomycosis. Verrucoid cutaneous lesion on the sternum due to disseminated disease

At 37°C: yeast phase (white to tan, glabrous to verrucoid)

Colony (microscopically)

At 25°C: usually sterile, hyaline, septate mycelia

At 37°C: thick-walled, broad-based, budding yeast

Histological staining

PAS: yeast stains pink to red
Silver: yeast stains dark blue to black
H & E: yeast stains weakly pale blue
Brown–Brenn: yeast stains dark blue

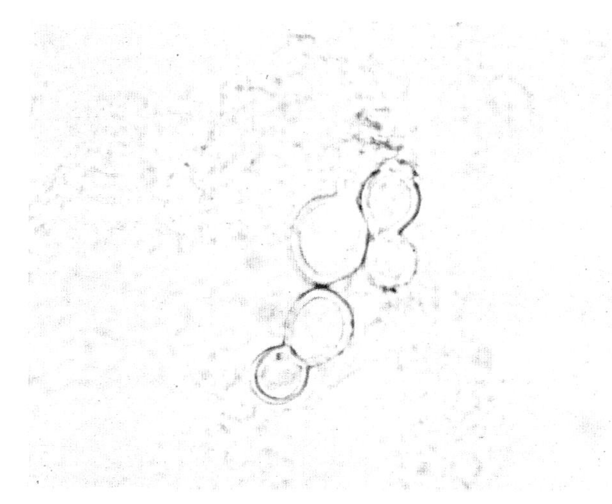

Figure 17.2 Microscopy of purulent drainage from a blastomycosis lesion shows broad-based budding yeast (KOH)

Coccidioides immitis

Source Endemic: southwestern parts of the USA, isolated areas of Central and South America; ecosystem: lower Sonoran life zone (dry soil: <10 inches of rainfall per year)

Clinical features Coccidioidomycosis: lung, skin, liver, bone, joints, urogenital tract, central nervous system

In vivo Large (60–80 μ) thick-walled spherules filled with endospores (5–7 μ)

In vitro

Colony (macroscopically)

At 25°C: mold phase (white to tan, cottony to granular; variable)

At 37°C: same as above

Colony (microscopically)

At 25°C: hyaline septate hyphae which separate into alternate arthroconidia

At 37°C: same as above

Histological staining

PAS: endospores stain pink to red
Silver: endospores stain black
H & E: endospores stain pale blue
Brown–Brenn: endospores stain dark blue

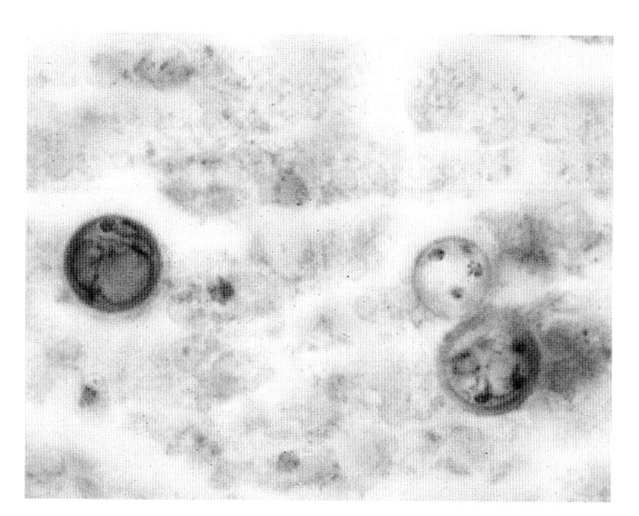

Figure 17.3 Microscopy of blastomycosis shows the double-contoured thick-walled yeast (PAS)

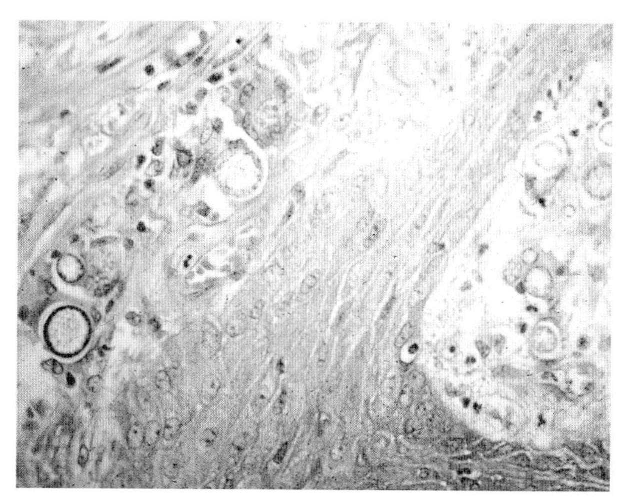

Figure 17.4 Histological section of coccidioidomycosis shows spherules at various stages of development (H & E)

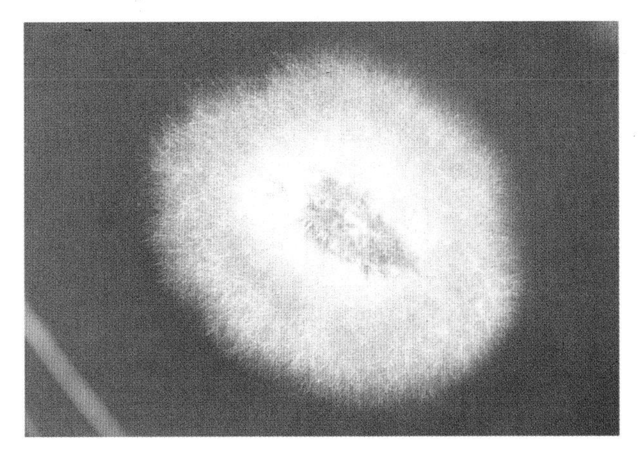

Figure 17.5 Obverse of *C. immitis* colony is white and cottony to granular when grown at 25°C

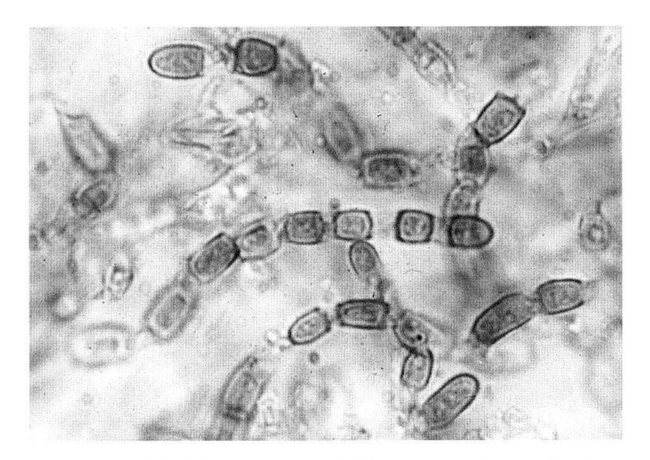

Figure 17.6 Microscopy of *C. immitis* shows hyphae which typically separate into alternating (highly infectious) arthroconidia (LPCB)

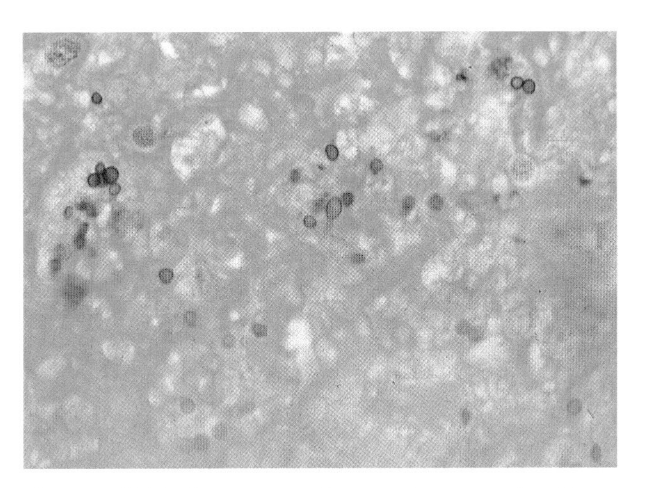

Figure 17.7 Microscopy of touch imprint of a histoplasmosis skin lesion reveals numerous small budding yeast (Gomori–methenamine–silver)

Histoplasma capsulatum

Source Endemic areas worldwide, primarily the Ohio and Mississippi River valleys of the USA; prefers soils enriched with high nitrogen material, especially bat and bird feces

Clinical features Histoplasmosis: lungs, organs of the reticuloendothelial system; skin lesions are variable, including papules, pustules, molluscum-like lesions and oral ulcerations

In vivo Small (2–5 μ) budding yeast which parasitizes histiocytes

In vitro

Colony (macroscopically)

At 25°C: mold phase (white or brown, cottony to granular)

At 37°C: yeast phase (white to tan, glabrous to crumbly)

Colony (microscopically)

At 25°C: hyphae producing pyriform microconidia and tuberculate macroconidia

At 37°C: small, single, budding, non-encapsulated yeast

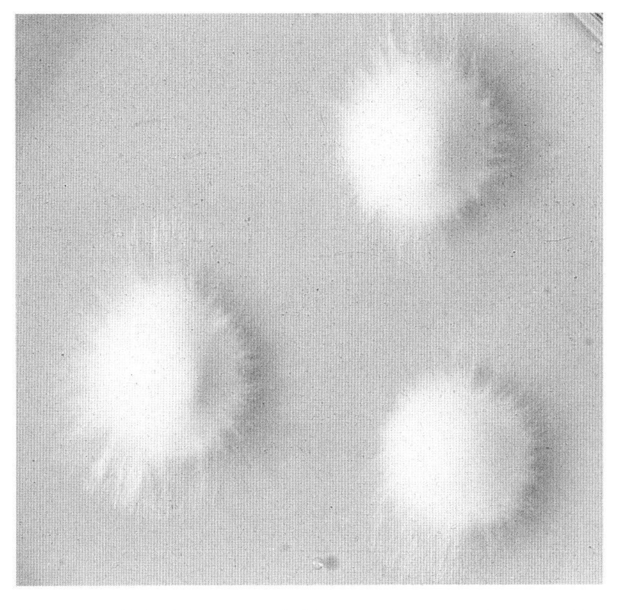

Figure 17.8 Obverse of *H. capsulatum* isolate *in vitro* is non-specific, white and cottony after 21 days of growth at room temperature (25°C) on SDA

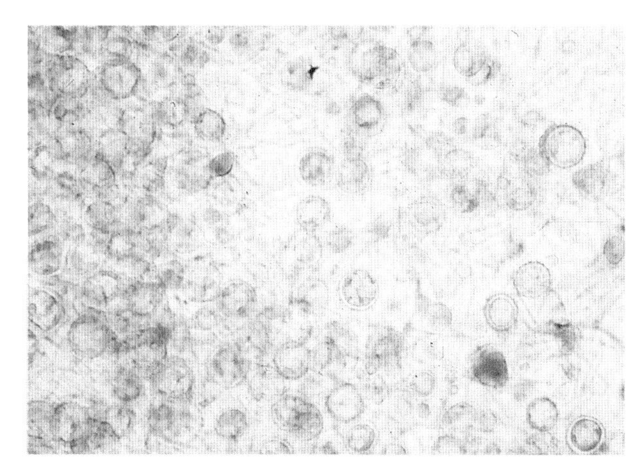

Figure 17.9 Microscopy of *H. capsulatum* shows septate mycelia and numerous thick-walled, tuberculate, macroconidia (LPCB)

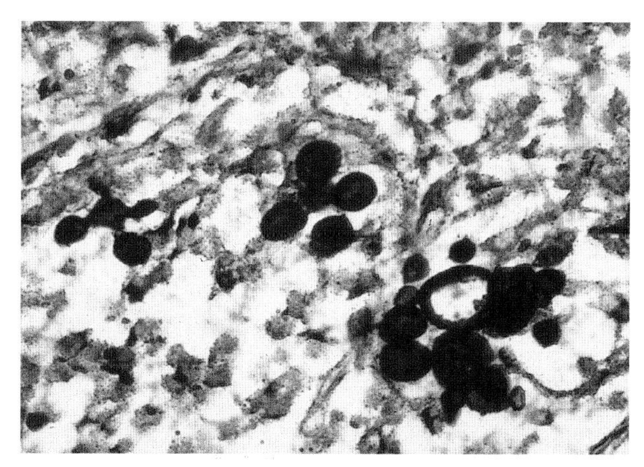

Figure 17.10 Biopsy of a paracoccidioidomycosis lesion shows a large parent cell giving rise to multiple dark-staining yeast (Gomori–methenamine–silver)

Histological staining

PAS: yeast stains pink to red with surrounding halo

Silver: yeast stains brown to black with surrounding halo

H&E: yeast stains weakly pale blue with surrounding halo

Brown–Brenn: yeast stains dark blue with surrounding halo

Paracoccidioides brasiliensis

Source Endemic: Central and South America; probable soil reservoir

Clinical features Paracoccidioidomycosis: lung, gastrointestinal mucosa, oral mucosa, lymphatics

In vivo Large (20–40 μ), non-encapsulated, multiple budding yeast (mariner's wheel)

In vitro

Colony (macroscopically)

At 25°C: mold phase (white to brown, glabrous to cottony)

At 37°C: yeast phase (white to tan, glabrous to crumbly)

Colony (microscopically)

At 25°C: hyaline septate hyphae with occasional globose conidia

Figure 17.11 Obverse of *P. brasiliensis* isolates *in vitro* is non-specific, white and cottony after 28 days of growth at room temperature (25°C) on SDA

At 37°C: large, non-encapsulated, multiple budding yeast

Histological staining

PAS: yeast stains pink to red

Silver: yeast stains brown to black

H&E: yeast stains weakly pale blue

Brown–Brenn: yeast stains dark blue

Sporothrix schenckii

Source Worldwide; soil and decaying plant material

Clinical features Sporotrichosis: subcutaneous with or without lymphatic involvement; chronic pulmonary; disseminated disease involving one or more organ systems

In vivo RARE small (2–6 μ), non-encapsulated, single budding yeast (cigar-shaped)

In vitro

Colony (macroscopically)

At 25°C: mold phase (gray becoming black, glabrous to velvety)

At 37°C: yeast phase (white to tan, glabrous)

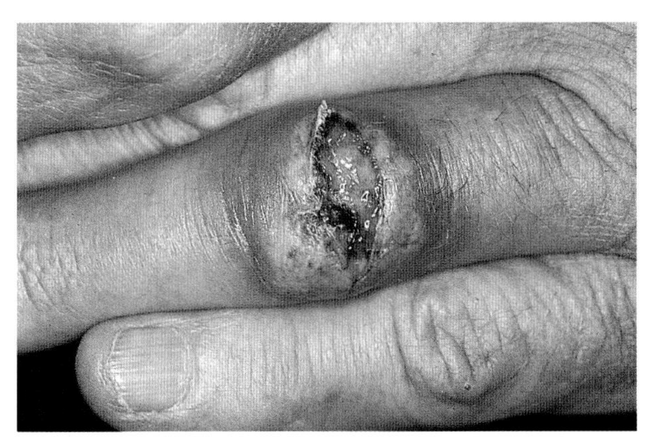

Figure 17.12 Ulcerated lesion of sporotrichosis on the hand at the site of primary inoculation

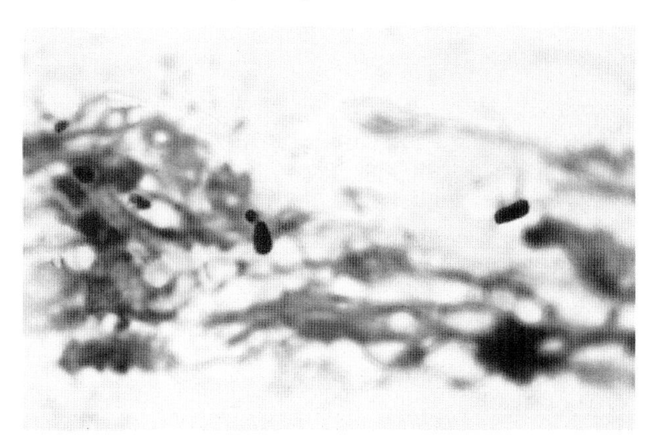

Figure 17.13 Microscopy of exudate from a sporotrichosis lesion smear shows dark-staining, cigar-shaped yeast (Gram's)

Colony (microscopically)

At 25°C: septate hyphae; conidiophores with ovoid conidia in a daisy petal arrangement

At 37°C: small, non-encapsulated, single budding yeast (some cigar-shaped)

Histological staining

PAS: yeast stains pink to red
Silver: yeast stains brown to black
H & E: yeast stains weakly pale blue
Brown–Brenn: yeast stains dark blue

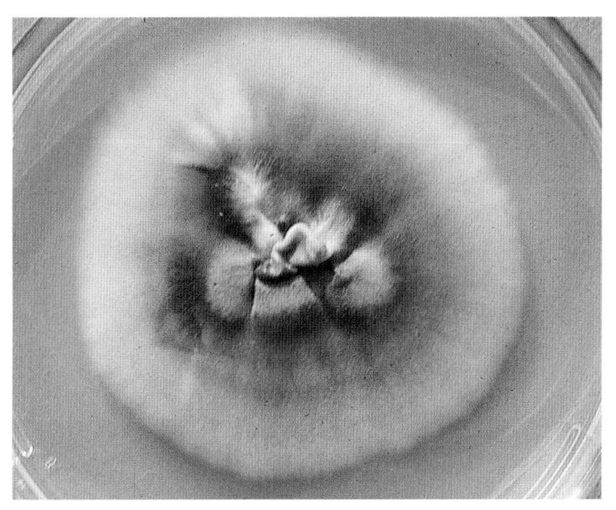

Figure 17.14 Obverse of *S. schenckii* isolate *in vitro* is dark and velvety after 7 days of growth at room temperature (25°C) on SDA

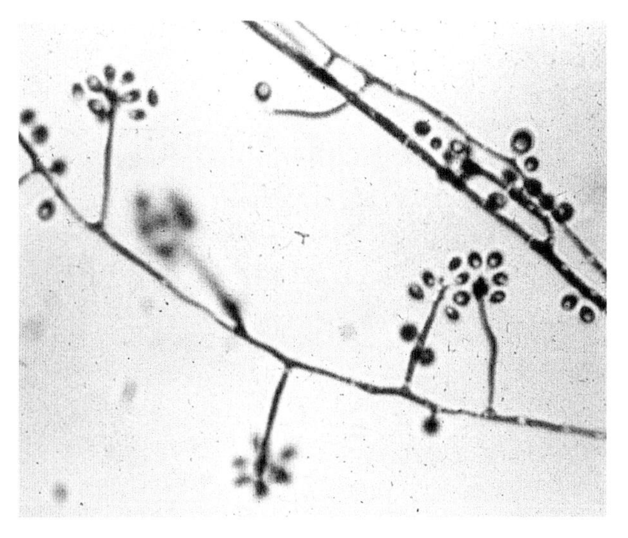

Figure 17.15 Microscopy of *S. schenckii* shows delicate septate mycelia, and conidiophores with ovoid conidia arranged in a daisy-like configuration

Section IV Viruses

by Philip R. Cohen

Chapter 18 Herpesviruses: Herpes simplex virus, varicella–zoster virus and cytomegalovirus 110

Chapter 19 Molluscum contagiosum 120

Chapter 18 Herpesviruses: Herpes simplex virus, varicella–zoster virus and cytomegalovirus

Human herpesviruses include herpes simplex virus (HSV) types 1 and 2, varicella–zoster virus (VZV), cytomegalovirus (CMV), Epstein–Barr virus (EBV), human herpesvirus 6 and human herpesvirus 7. HSV and VZV infections of the mucous membranes and skin are common. Several laboratory diagnostic methods are available for the diagnosis, differentiation and subtyping of HSV and VZV infections. A Tzanck smear preparation is an inexpensive, rapid, morphological technique for confirming a suspected diagnosis of HSV or VZV infection in the office. Other techniques that may be performed in the office to diagnose either a HSV or VZV infection include viral culture and biopsy of the lesion. However, these methods require additional time for processing the specimen before the diagnosis of a viral infection can be confirmed.

In contrast to HSV and VZV infections, CMV infection rarely involves the skin. Also, whereas HSV and VZV only have intranuclear viral inclusions, CMV may have either intranuclear or intracytoplasmic inclusions, or both. Confirmation of a suspected cutaneous CMV infection cannot be promptly established in the office laboratory. However, a lesional biopsy may be performed in the office to evaluate the tissue specimen for characteristic pathological changes.

Herpes simplex and varicella–zoster virus infections

Clinical presentation

Cutaneous lesions of an HSV infection classically appear as erythematous-based grouped vesicles with erythema of the adjacent skin (Figures 18.1 and 18.2). Clinical evidence of a recurrent HSV infection may be the presence of focal areas of macular postinflammatory hyperpigmentation. The clinical presentation of VZV infection is variable, usually appearing as erythematous-based vesicles in a dermatomal distribution (Figures 18.3 and 18.4). However, disseminated VZV infection presents with multiple individual vesicles with or without additional dermatomal lesions.

Occasionally, an HSV infection may mimic a VZV infection when the individual lesions are located in a dermatome. Less often, a 'localized' VZV infection morphologically mimics HSV infection. The clinical presentations of both disseminated VZV infection without dermatomal lesions and disseminated HSV infection are identical.

Tzanck smear preparation

Tzanck smear preparations (Tables 18.1 and 18.2) are easy to perform not only in the office, but also at the bedside of a hospitalized patient (Figures 18.5–18.15). Although the technique enables confirmation of a suspected HSV or VZV infection, it is not able to distinguish either (1) the subtype of HSV infection, or (2) between HSV and VZV infection.

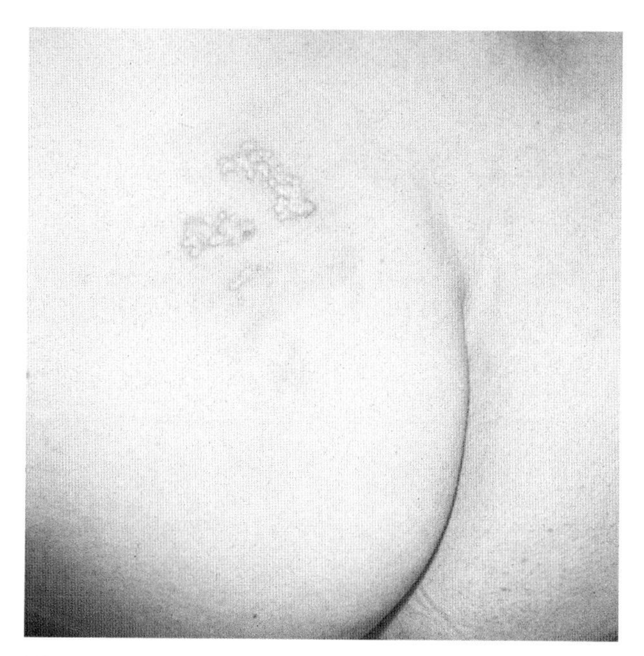

Figure 18.1 Buttocks of a 65-year-old woman with recurrent asymptomatic HSV infection. Reproduced with permission of Cahners Publishing Co., a Division of Reed Publishing USA, New York, NY; from Cohen & Young Jr, 1988

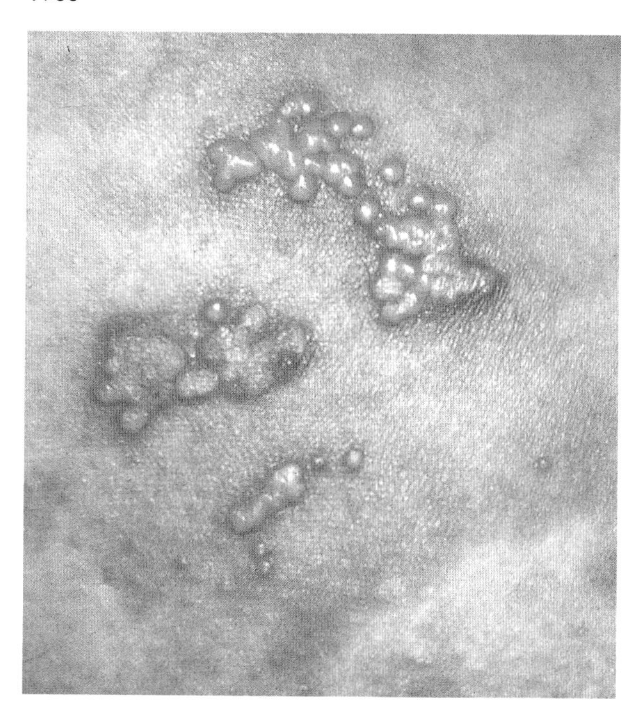

Figure 18.2 HSV type 2 infection on the left buttock shows grouped vesicles with surrounding erythema. Reproduced with permission of Cahners Publishing Co., a Division of Reed Publishing USA, New York, NY; from Cohen & Young Jr, 1988

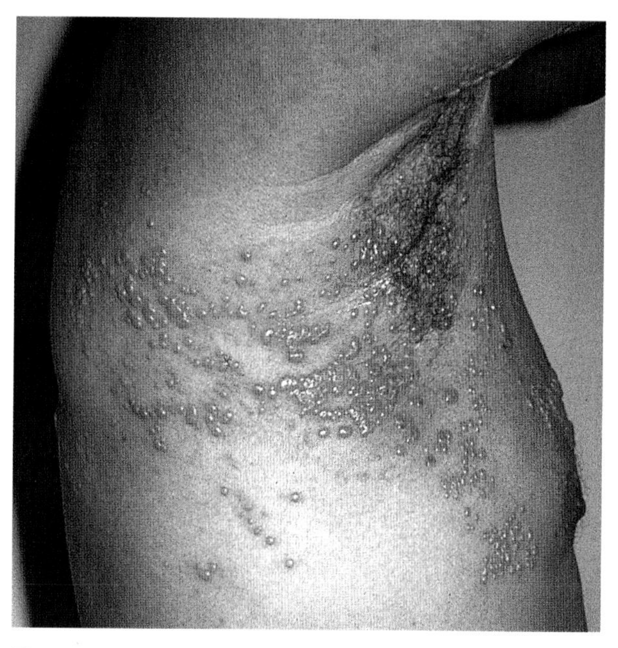

Figure 18.3 Grouped vesicles with an erythematous base involving dermatomes T3 and T4 on the right flank, chest and back of a middle-aged man due to VZV infection. Reproduced with permission of W.B. Saunders Co., Philadelphia, PA; from Cohen, 1994

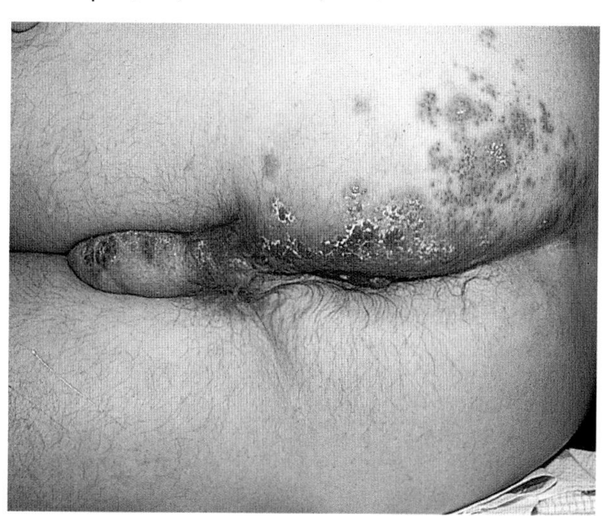

Figure 18.4 This patient was receiving chemotherapy for acute lymphocytic leukemia and developed a VZV infection on the left buttock and scrotum (dermatome S2). There are erythematous-based vesicles and crusted erosions

The Tzanck smear preparation is less accurate than viral culture in detecting HSV infection. However, the Tzanck smear provides a more immediate answer and a higher positive yield than viral cul-

Table 18.1 Tzanck smear technique

1. Select a lesion. The yield for obtaining a positive smear is highest from a new vesicle, followed by a pustule, and then a crusted ulcer
2. Wipe the lesion with alcohol, and allow to dry for 1 min
3. Using a number 15 scalpel blade, either remove the crust or unroof the vesicle or pustule
4. Holding the number 15 scalpel blade at an angle $<90°$, vigorously scrape the lesion base with the edge of the blade
5. Gently transfer the material from the blade to a glass slide by repeatedly and gently touching the blade to the slide (forceful smearing will grind the delicate cells, resulting in streaked and crushed nuclei on the stained preparation)
6. Allow the smear to air-dry[a] (may first spray the smear with Cytoprep® to preserve cellular detail)
7. Flood the slide with staining solution for 30–60 s. Use either Giemsa, Wright's, methylene or toluidine blue stains, or Sedi-Stain®
8. Rinse the excess stain off the slide with tap water and allow the slide to air-dry. It is not necessary to heat the slide, which may be detrimental if overdone
9. To ensure maximum optical resolution, apply 1–2 drops of mounting medium for permanent preparations, or either immersion oil or tap water for temporary preparations, then place a coverslip over the slide
10. To arrange the microscope, rack the substage condenser up towards the stage, and adjust the field diaphragm for optimal illumination and resolution
11. Scan and focus the smear, using the ×10 and ×40 objectives, respectively. The ×100 oil objective may then be used to confirm the characteristic cytological features of HSV infection, namely:

    Multinucleated giant cells with molded-together nuclear contours and homogeneously stained ground-glass nuclear chromatin;

    Intranuclear viral inclusion bodies (Lipschütz or Cowdry; seen only with stains containing eosin, such as Papanicolaou's stain)

[a] If Papanicolaou's stain is to be used, the smear must be fixed immediately in alcohol

Reproduced with permission of Cahners Publishing Co., a Division of Reed Publishing USA, New York, NY; from Cohen & Young Jr, 1988

Table 18.2 Reagents used for diagnosing viral infections in the office laboratory

PMS fungal / Tzanck stain[a]
Delasco
Catalog no. 34103
Sizes: 0.5 oz and 1.0 oz bottles
Dermatologic Lab & Supply Co., Inc.
608 13th Avenue, Council Bluffs, IO 51501
Tel: 1-800-831-6273
1-888-DELASCO
(712)-323-3269
Fax: (712)-323-1156
E-mail: questions@delasco.com

Sedi-Stain®[b]
Curtin Matheson Scientific
Catalog no. 23 590-992
Size: 12.5 ml bottle
Now owned by Fisher Scientific Co.
Home Office: Park Lane, Pittsburgh, PA
Tel: 1-800-640-0640
1-888-437-0535
Curtin Matheson Scientific is a distributor for Sedi-Stain, produced by BD Primary Care, Sports, MD
Tel: (410)-316-3300

[a]Contains toluidine blue and basic fuchsin in 30% alcohol; [b]Contains crystal violet (0.10%), safranin O (0.25%), ammonium oxalate (0.03%), ethyl alcohol (SD-3A), water and stabilizers (89.62%)

ture for VZV infection; this is secondary to the increased difficulty in isolating VZV in culture.

A positive smear is more readily obtained from a new vesicle, followed by a pustule. A crusted ulcer

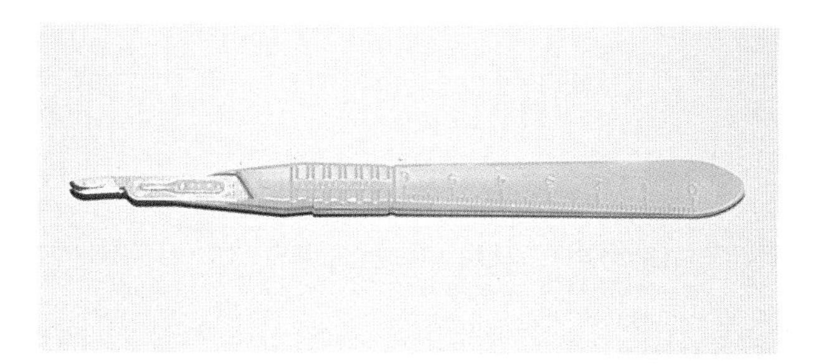

Figure 18.5 A number 15 Bard–Parker scalpel blade is used during Tzanck smear preparation

Figure 18.6 A glass microscope slide is used for the Tzanck smear preparation

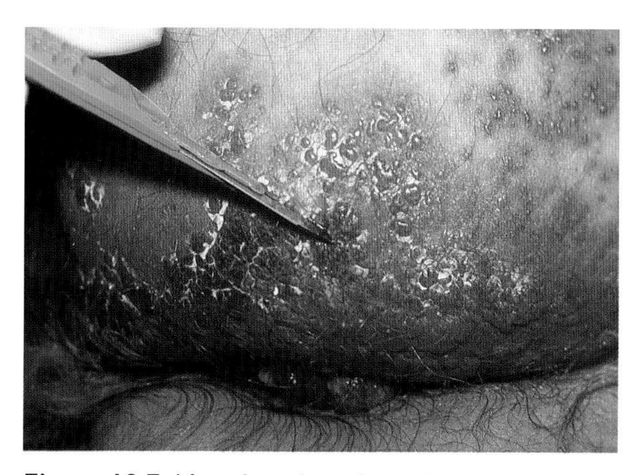

Figure 18.7 After the selected vesicle has been wiped with alcohol and allowed to dry for 1 min, the scalpel blade is used to unroof the lesion and to vigorously scrape the base of the vesicle

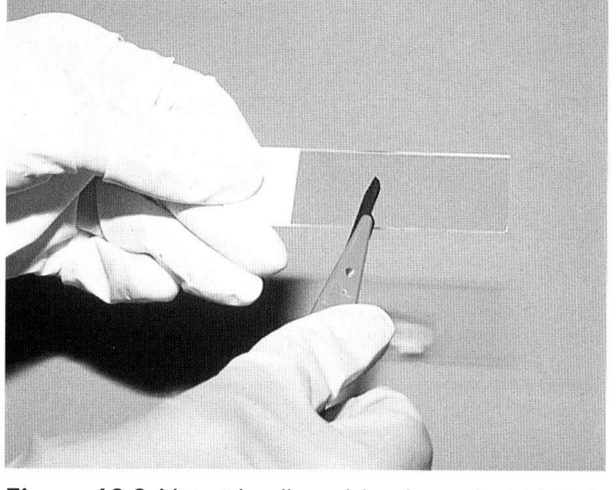

Figure 18.8 Material collected by the scalpel blade is gently transferred to a glass slide and allowed to air-dry

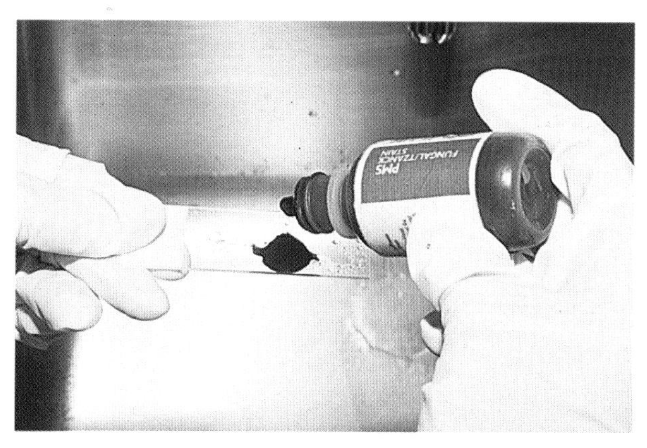

Figure 18.9 The glass slide is flooded with staining solution (PMS fungal / Tzanck)

Figure 18.10 The staining solution remains on the slide for 30–60 s

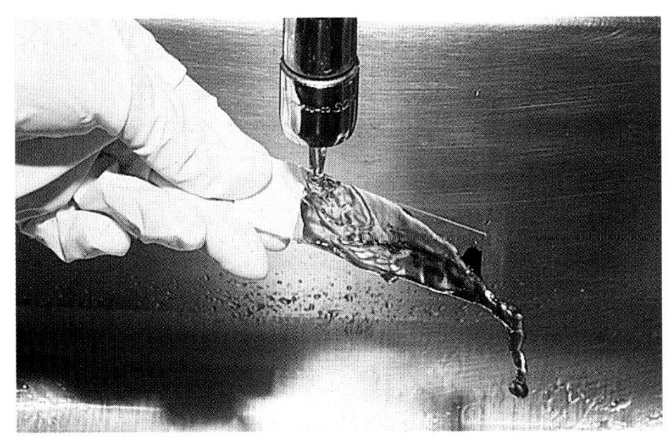

Figure 18.11 The excess staining solution is gently rinsed off the slide with tap water; the water should make contact with an unstained part of the slide to avoid washing the stained sample off the slide

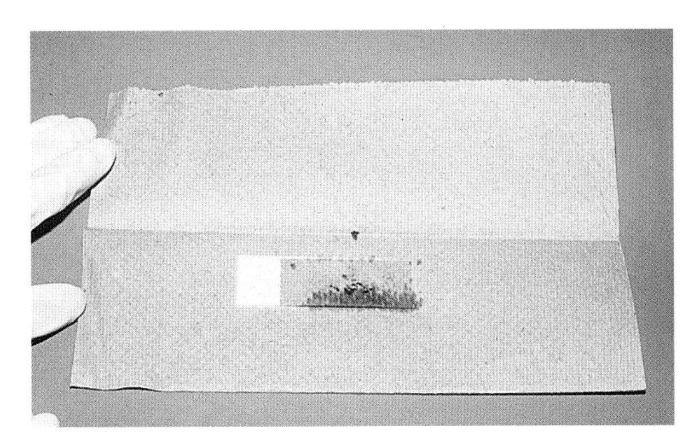

Figure 18.12 The slide is allowed to air-dry. Drying may be hastened by placing the slide on a paper towel

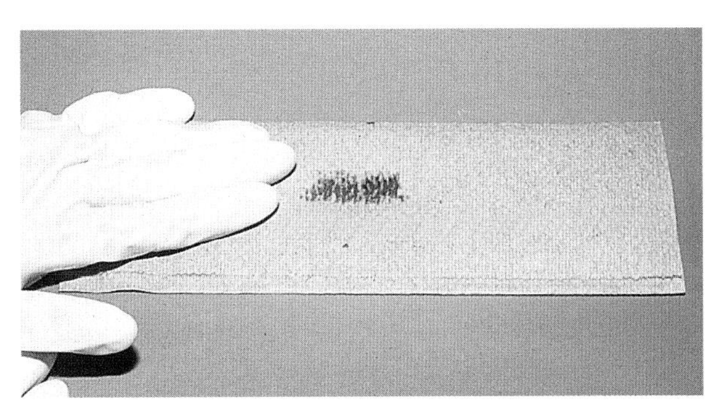

Figure 18.13 The paper towel is carefully folded on top of the slide and VERY gently pressed against the slide

has the lowest yield for providing a positive smear. When performing a Tzanck smear preparation from a crusted lesion, first remove the crust, then make the smear from scrapings of the base of the lesion under the crust.

Figure 18.14 A drop of oil and then a coverslip may be applied to optimize optical resolution on microscopy

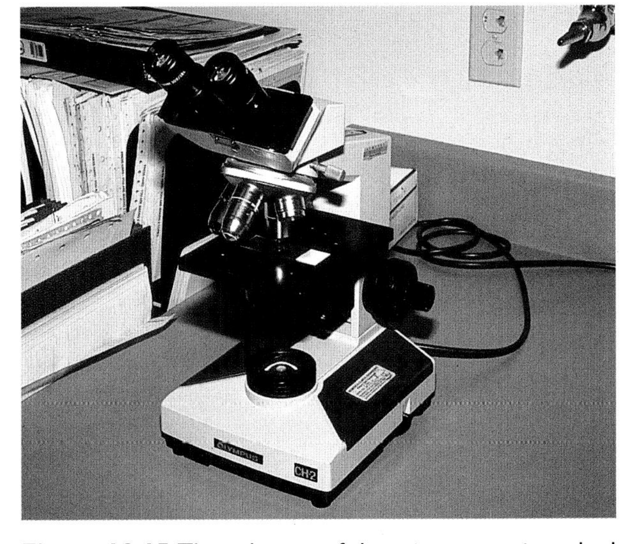

Figure 18.15 The substage of the microscope is racked up towards the stage, and the field diaphragm adjusted for optimal illumination and resolution

After staining, examine the slide under ×40 magnification with a light microscope. A slightly enlarged nucleus and a thickened nuclear membrane are observed in recently infected cells. Cell fusion and typical multinucleated giant cells are observed as the infection progresses. The pathognomonic findings of herpesvirus infection on Tzanck smear preparation are: (1) Multinucleated giant cell formation; (2) a peculiar ground-glass appearance (or homogenization) of the nuclear chromatin; and (3) molded-together or 'faceted' nuclear contours (Figures 18.16–18.18).

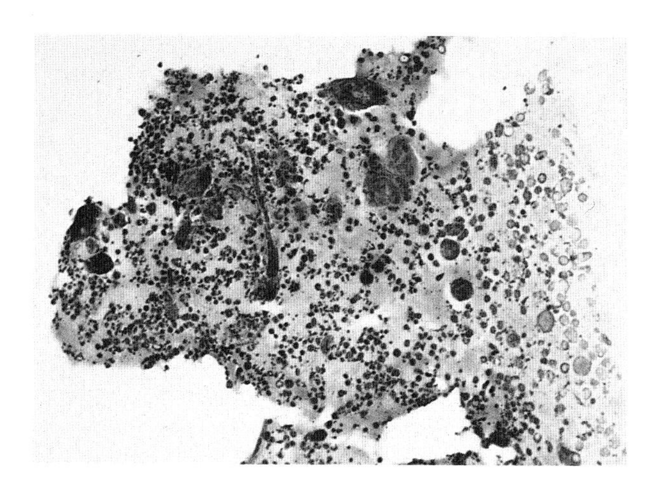

Figure 18.16 The smear is initially scanned at lower magnification. Several multinucleated giant cells can be seen [toluidine blue and basic fuchsin (PMS fungal / Tzanck); × 50]

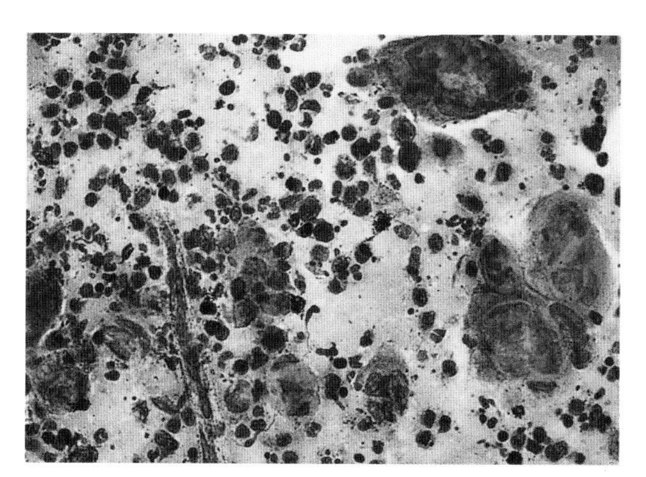

Figure 18.17 Characteristic multinucleated giant cells of a herpesvirus infection are evident (PMS fungal / Tzanck; × 100)

Viral culture

Viral tissue culture is the 'gold standard' for establishing the diagnosis of HSV infection, and requires the use of viral transport medium (Figure 18.19) and a plastic applicator (Figure 18.20). Viral culture for HSV is rapid. Within 24–48 h after inoculation, >50% of HSV isolates demonstrate a cytopathic effect and, within 3–4 days, this proportion is >90%. For VZV, it may take from 7 days to 3 weeks to produce the characteristic cytopathic changes because VZV proliferates at a slower rate *in vitro* than HSV.

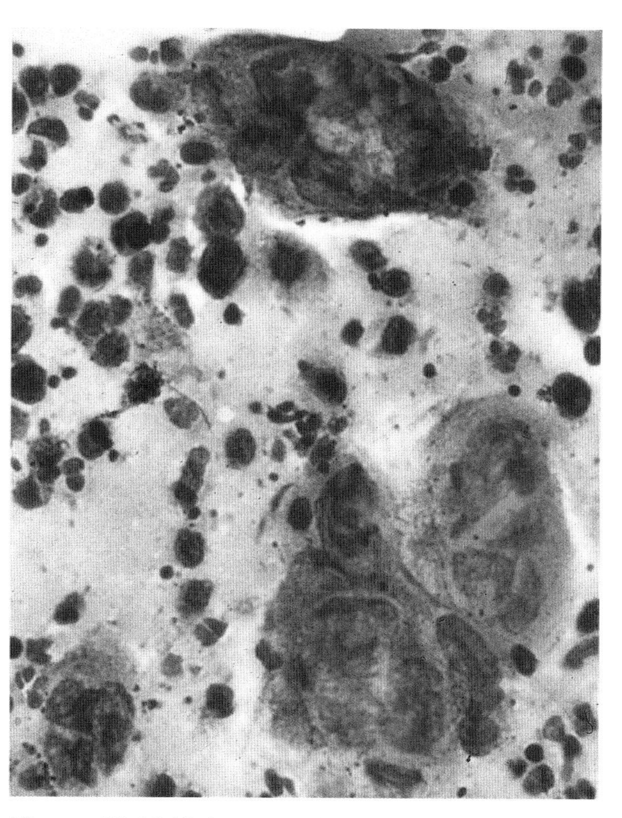

Figure 18.18 Higher magnification confirms the cytological features of herpesvirus infection: Multinucleated giant cells with molded-together nuclear contours and homogeneously stained, ground-glass, nuclear chromatin (PMS fungal / Tzanck; × 150)

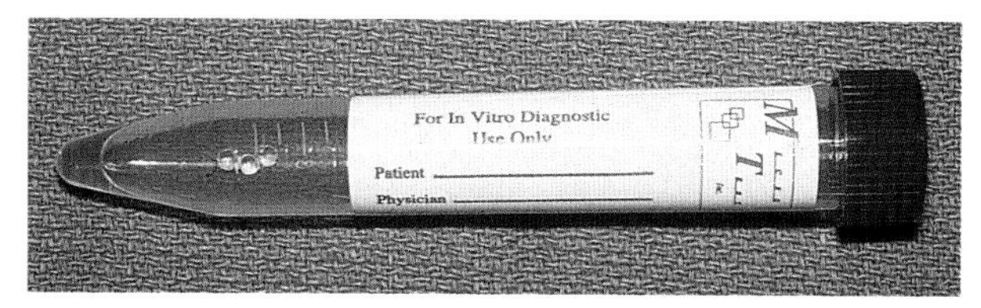

Figure 18.19 Viral transport medium contains a buffered salt solution with protein and antibiotics. This solution should be stored at 4°C (in a refrigerator) until ready for use

Figure 18.20 A **plastic** applicator with a **dacron** (or rayon) tip should be used to obtain the sample to be sent for viral culture

Figure 18.21 A **wood** applicator should NEVER be used to obtain a sample for viral culture

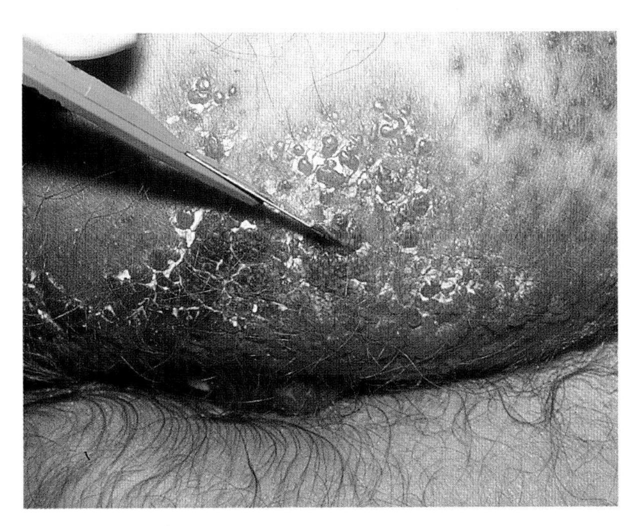

Figure 18.22 After the selected vesicle has been wiped with alcohol and allowed to dry for 1 min, a number 15 scalpel blade is used to unroof the lesion

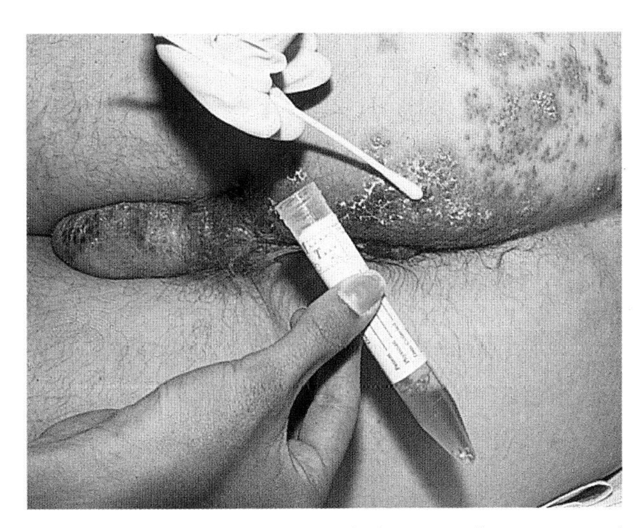

Figure 18.23 A dacron-tipped plastic applicator is used to absorb the vesicle fluid for viral culture. If fluid is not available, the applicator tip may be rubbed against the base of the lesion

When the possibility of HSV or VZV infection is entertained and a viral culture is to be performed, it is important that the specimen be appropriately collected and transported. The requirements for successful tissue culture of virus include: (1) Inoculation of specimens containing live virus; (2) proper care in obtaining, transporting, and inoculating specimens; and (3) prevention of bacterial or fungal contamination.

Virus isolation of VZV is more difficult than isolation of HSV as VZV is less tolerant of transport and / or storage conditions than HSV.

Isolation of HSV is more frequent when the specimen is taken from either an unroofed vesicle or shallow ulcer compared with either a crusted lesion or a maculopapular area. After the vesicle has been unroofed, the vesicular fluid and cellular debris should be swabbed with a **dacron**-tipped wire or **plastic** applicator (as wood and cotton may contain substances that can kill the virus; Figure 18.21). If the lesion is crusted, the necrotic debris should first be removed by washing with sterile saline before obtaining cells from the lesion base for culture (Figures 18.22 and 18.23).

Calcium alginate swabs (used for culturing *Neisseria gonorrhoeae*) inactivate HSV and may therefore give a false-negative result for the presence of virus. The purulent material from a pustular lesion or an insufficient number of parabasal cells containing replicating virus (insufficient for detection based on the sensitivity of the assay) may also

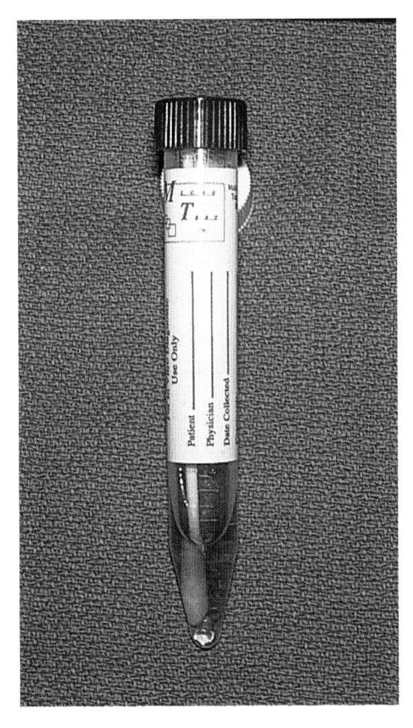

Figure 18.24 The applicator tip is placed in the viral transport medium; the excess applicator is broken off and the cap replaced

Figure 18.25 The specimen should be kept at 4°C (for example, in a cup of ice) until it arrives at the laboratory as there is a marked reduction of virus titers when the specimen is stored in transport medium at room temperature

give a false-negative result. Avoidance of contamination of the specimen with alcohol, blood, soap or stool is also important as these substances may either inactivate HSV or be toxic to the cell culture.

Viral transport media contain a buffered salt solution with protein and antibiotics. If the anticipated transport time is >8 h, the swabs should be removed after vigorous swirling in the media. At 4°C (refrigerator or ice bath), HSV remains stable in most viral transport media until inoculation into tissue culture (Figures 18.24 and 18.25). If viral transport medium is not available, the culture swab or tissue specimen should be kept moist, using non-bacteriostatic sterile saline. Destruction of the viral lipid membrane and inactivation of the virus will occur if the specimen is permitted to dry out.

When a lesional biopsy has been performed and the tissue specimen is received by the laboratory, it is ground by a tissue homogenizer into a 10% suspension in tissue culture medium. After centrifugation at 1500 rpm for 10 min (to remove cellular material and debris), a small volume of the supernatant (approximately 0.25 ml) is directly inoculated into the tissue culture.

Fibroblast cells [maintained at 37°C in Eagle's minimal essential medium supplemented with amino acids (1% L-glutamine), antibiotics, antifungals and fetal calf serum (2–5%)] are often used for inoculation of the specimen. These cells not only give an easily recognizable cytopathic effect, but also permit the growth of other viruses that may be present in vesicular specimens. Typical HSV-induced cytopathic effects may be observed using a light microscope at lower magnifications (×10–40). Initially, there is cytoplasmic granulation and cell rounding; subsequently, the cells become swollen and refractile before eventually detaching from the culture tube. The cytopathic effects induced by HSV differ from those due to VZV.

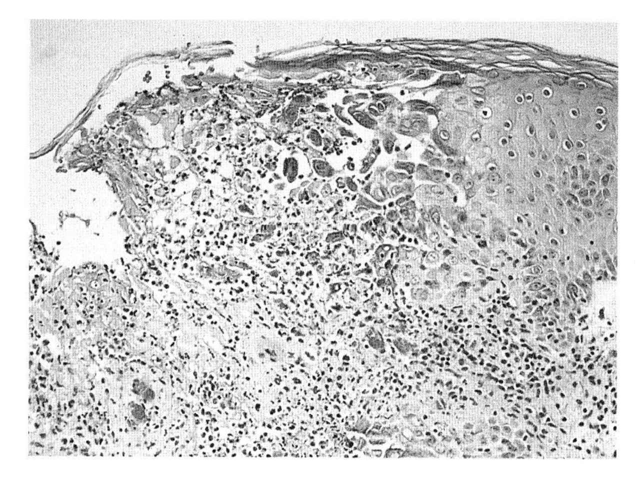

Figure 18.26 Low magnification of a skin biopsy of recurrent HSV infection shows an ulcer border with balloon cells and multinucleated giant cells. There is papillary dermal edema and a mixed inflammatory infiltrate of lymphocytes and neutrophils in the dermis. Reproduced with permission of W.B. Saunders Co., Philadelphia, PA; from Cohen, 1994 (H & E; $\times$50)

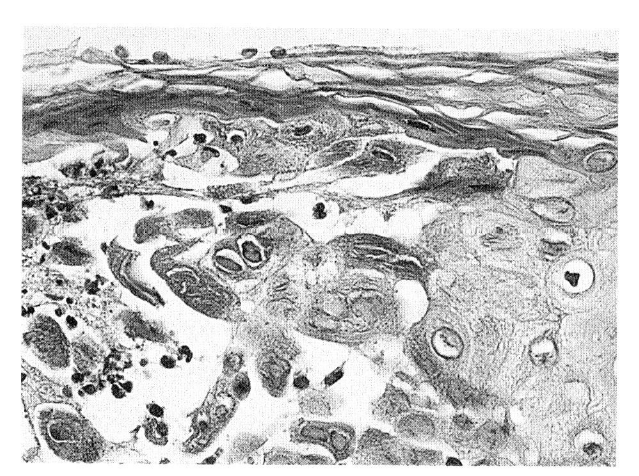

Figure 18.27 Higher magnification of the skin biopsy in Figure 18.26 shows multinucleated giant cells at the edge of the ulcer. Reproduced with permission of W.B. Saunders Co., Philadelphia, PA; from Cohen, 1994 (H & E; $\times$100)

Biopsy

A biopsy taken from the edge of an unusual lesion may help to establish the diagnosis of herpetic infection. In addition to the microscopic examination of the surface epidermis, careful evaluation of hair follicles should be performed as early and characteristic changes of herpetic infection may occur at this site. Ballooned or multinucleated keratinocytes and typical nuclear changes of viral infection are present on a positive biopsy (Figures 18.26 and 18.27).

Cytomegalovirus

Clinical presentation

CMV infection of the skin is uncommon. Disseminated CMV infection is present in most patients with virus-associated cutaneous lesions. Although CMV-related skin lesions (typically a rubelliform-type exanthem) may appear in otherwise healthy individuals, many of the patients with cutaneous CMV infection have an associated underlying disorder such as burns, cancer, human immunodeficiency virus (HIV) infection or iatrogenic immunosuppression (following organ transplantation).

The lesions of cutaneous CMV infection are variable in morphology. They include dermatitis (diaper), epidermolysis, granulation tissue (burns), morbilliform eruption, papules, petechiae, purpura, pyoderma, ulcerations, urticaria, verrucous nodules and plaques, and vesicles. Perianal ulcers are the most specific cutaneous manifestation of CMV infection.

Viral culture

Diagnosis of CMV infection may be confirmed by viral culture. The traditional sources for culture are the blood, throat and urine; confirmation of viremia from a buffy coat culture is an excellent indicator of active systemic CMV infection. Tissue obtained from a biopsy of a suspected CMV skin lesion may also be cultured.

CMV only grows in fibroblast culture. Within 1–2 weeks, CMV-induced cytopathic effects may be noted. However, if the viral titers are low, these changes may be delayed for as long as 6 weeks.

Biopsy

Cutaneous CMV infection has a predilection for

the endothelial cells of veins within the dermis. The term 'owl-eyes inclusions' refers to the characteristic basophilic intranuclear inclusions that are surrounded by a clear halo. These intracytoplasmic inclusions are perinuclear and appear as granular, periodic acid–Schiff (PAS)-positive structures. Immunohistochemical testing, using CMV-specific monoclonal antibodies, of the tissue specimen may confirm the diagnosis of CMV infection.

Bibliography

Cohen PR. Tests for detecting herpes simplex virus and varicella–zoster virus infections. *Dermatol Clin* 1994; 12:51–68

Cohen PR, Grossman ME. Clinical features of human immunodeficiency virus-associated disseminated herpes zoster virus infection—a review of the literature. *Clin Exp Dermatol* 1989;14:273–6

Cohen PR, Kazi S, Grossman ME. Herpetic geometric glossitis: A distinctive pattern of lingual herpes simplex virus infection. *South Med J* 1995;88:1231–5

Cohen PR, Young AW Jr. Herpes simplex: Update on diagnosis and management of genital herpes infection. *Med Aspects Hum Sexual* 1988;22:93–100

Lesher JL Jr. Cytomegalovirus infections and the skin. *J Am Acad Dermatol* 1988;18:1333–8

Chapter 19 Molluscum contagiosum

Molluscum contagiosum is a poxvirus infection which may present with cutaneous or mucosal lesions, or both. Molluscum bodies are large brick-shaped intracytoplasmic viral particles present within the keratinocytes. A molluscum preparation is a simple and rapid procedure that is easily performed in the office to confirm a suspected diagnosis of molluscum contagiosum. Alternatively, biopsy of the lesion for microscopic examination may also establish the diagnosis.

Clinical presentation

The lesions of molluscum contagiosum are small (typically 1–4 mm), flesh-colored to pearly, dome-shaped papules that may be found on any area of the skin or mucosa (Figures 19.1 and 19.2). Typically, there is central umbilication. The white curd-like core contains the intracytoplasmic inclusions (molluscum bodies or Henderson–Patterson bodies) which can readily be expressed.

Although the characteristic morphology of molluscum contagiosum is distinctive, the cutaneous lesions may mimic other conditions when the clinical presentation is not classical. The clinical differential diagnoses include acrochordons, basal cell carcinoma, lichen planus, pyoderma, varicella and verruca. In patients with HIV infection, cutaneous lesions of disseminated cryptococcosis may be identical in appearance to those of molluscum contagiosum.

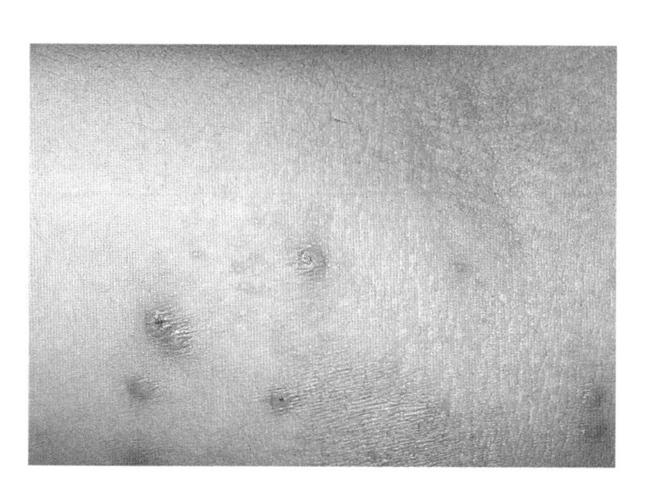

Figure 19.1 Antecubital fossa of a 12-year-old girl presents several flesh-colored umbilicated papular lesions of molluscum contagiosum

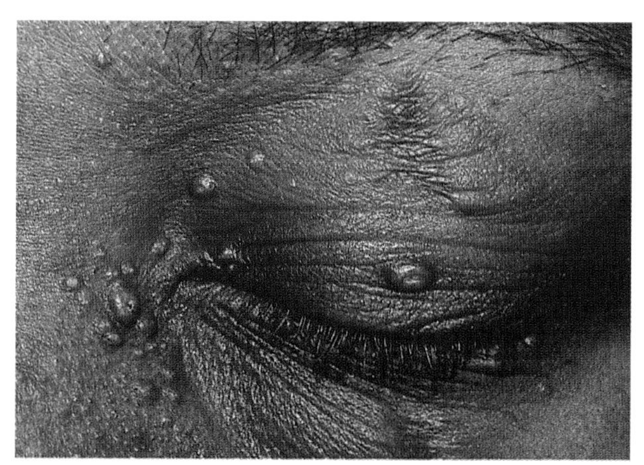

Figure 19.2 Periocular lesions of molluscum contagiosum in a 21-year-old woman with HIV infection

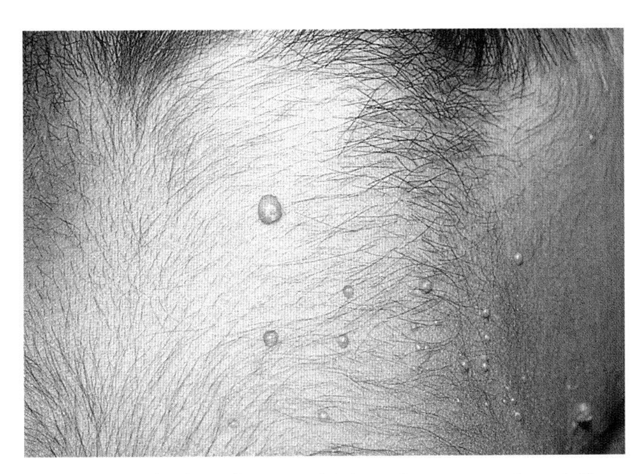

Figure 19.3 This 8-year-old boy has several umbilicated papules of varying size on the right posterior neck. As a diagnosis of molluscum contagiosum is suspected, a molluscum preparation is to be performed on the larger, centrally located, lesion

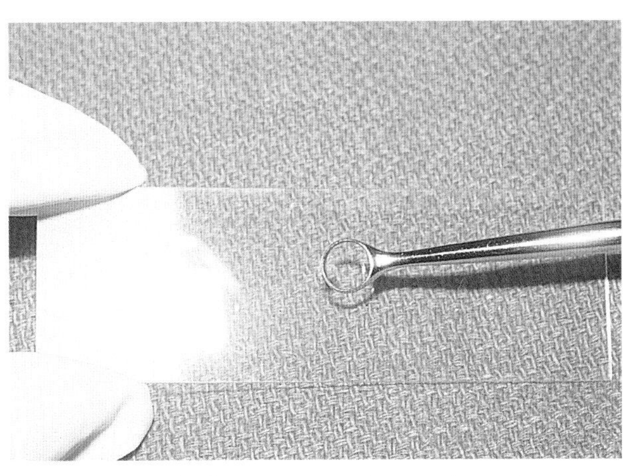

Figure 19.5 The lesion is placed on a glass microscope slide

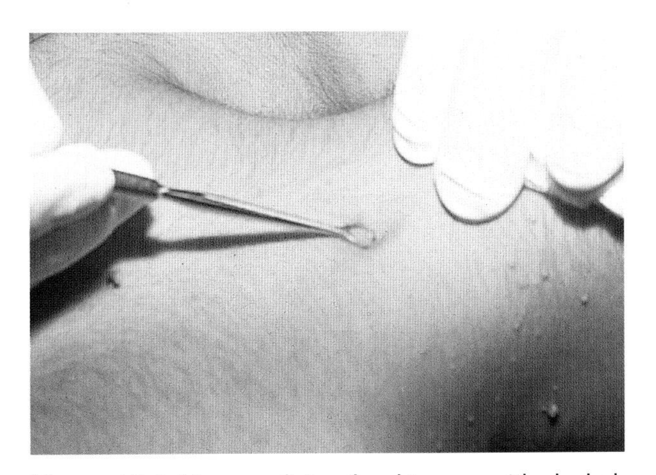

Figure 19.4 After sterilizing the skin area with alcohol, a curette is used to remove the lesion

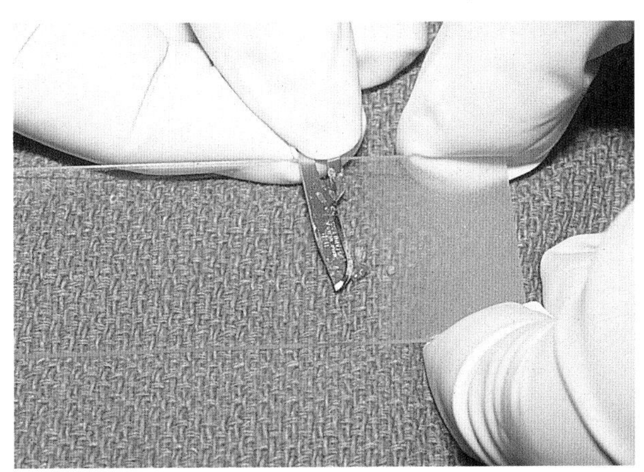

Figure 19.6 A number 15 scalpel blade is used to squash the lesion

Molluscum preparation

This technique is similar to the Tzanck smear preparation and enables rapid confirmation in the office laboratory of the clinically suspected diagnosis of molluscum contagiosum (Figure 19.3). The skin lesion is removed with either a curette or a scalpel with a number 15 blade (Figure 19.4). Alternatively, the central core of the lesion may be expressed, using a double-ended Schamberg comedone extractor after a minimal incision of the papule has been made. The lesion (or the pearl-like central core) is squashed and firmly spread onto a glass micro-

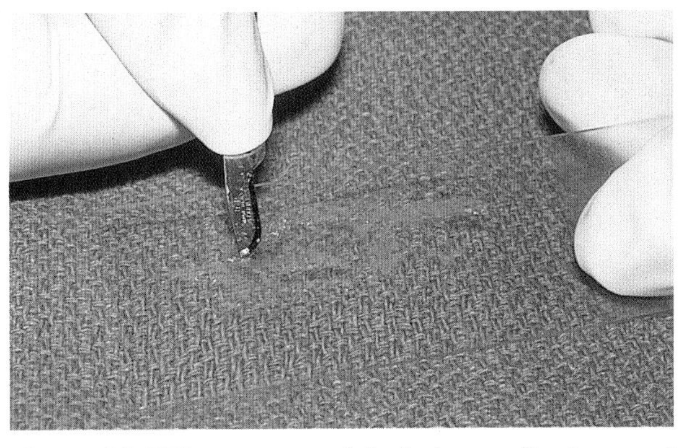

Figure 19.7 The contents of the lesion are firmly spread onto the glass microscope slide

scope slide; this may be carried out by using either the number 15 scalpel blade or another glass slide (Figures 19.5–19.7).

Any of several different stains may be used to visualize the molluscum bodies: Giemsa, Gram's, H&E, Wright's, methylene blue, Papanicolaou's or toluidine blue. Commercially available agents for staining molluscum preparations also include PMS fungal / Tzanck stain and Sedi-Stain® (see Table 18.2, page 112). A few drops of stain, sufficient to cover the slide, are added. After 30–60 s, the excess stain is gently washed off with tap water. A coverslip is placed over the stained tissue and the slide is examined using the light microscope.

Among the keratinocytes, numerous amorphous, darkly staining, smaller molluscum bodies may be observed (Figures 19.8–19.11). Although the clinical lesions of cryptococcosis in HIV-infected patients may resemble those of molluscum contagiosum, the molluscum bodies demonstrated by a molluscum preparation are morphologically distinct from *Cryptococcus neoformans* organisms (Figure 19.12).

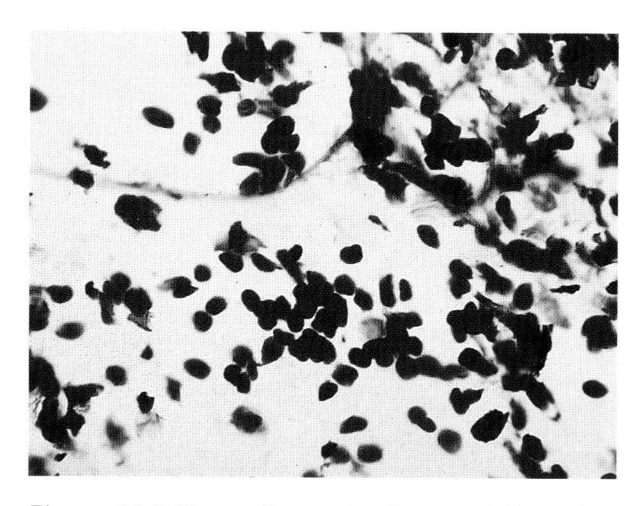

Figure 19.8 The molluscum bodies are visible at low magnification (PMS fungal / Tzanck; × 50)

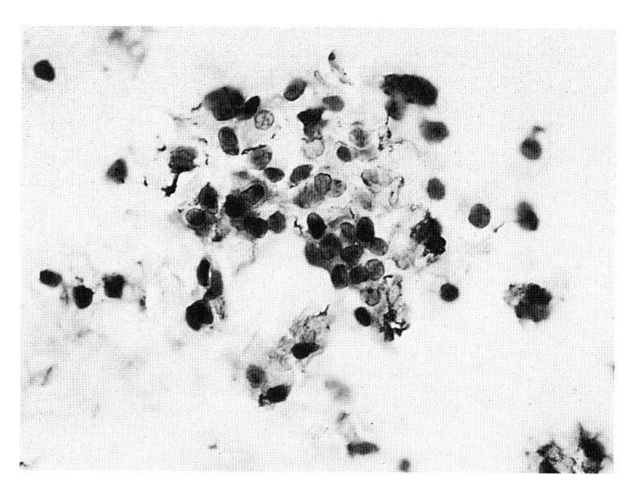

Figure 19.10 Numerous amorphous, darkly staining, molluscum bodies can be identified at low magnification (Giemsa; × 50)

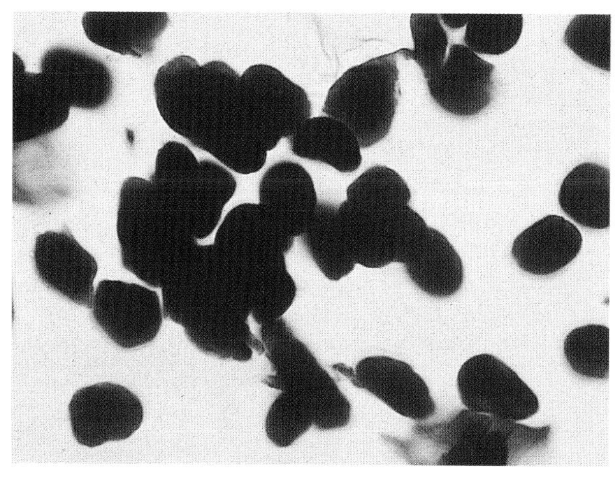

Figure 19.9 The amorphous viral inclusion bodies can be seen at higher magnification (PMS fungal / Tzanck; × 100)

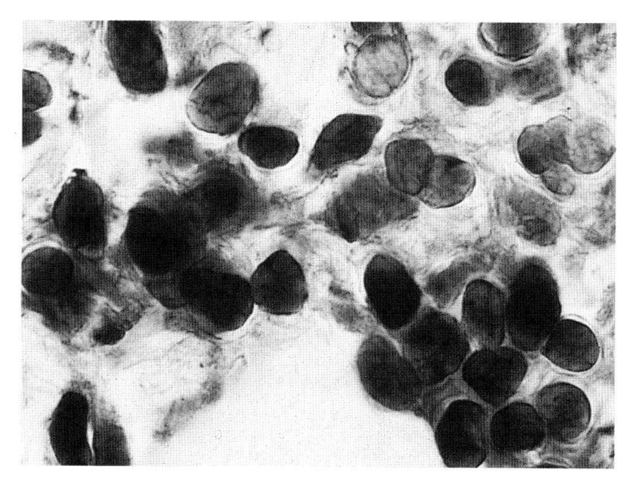

Figure 19.11 Higher magnification reveals the viral inclusion bodies (Giemsa; × 100)

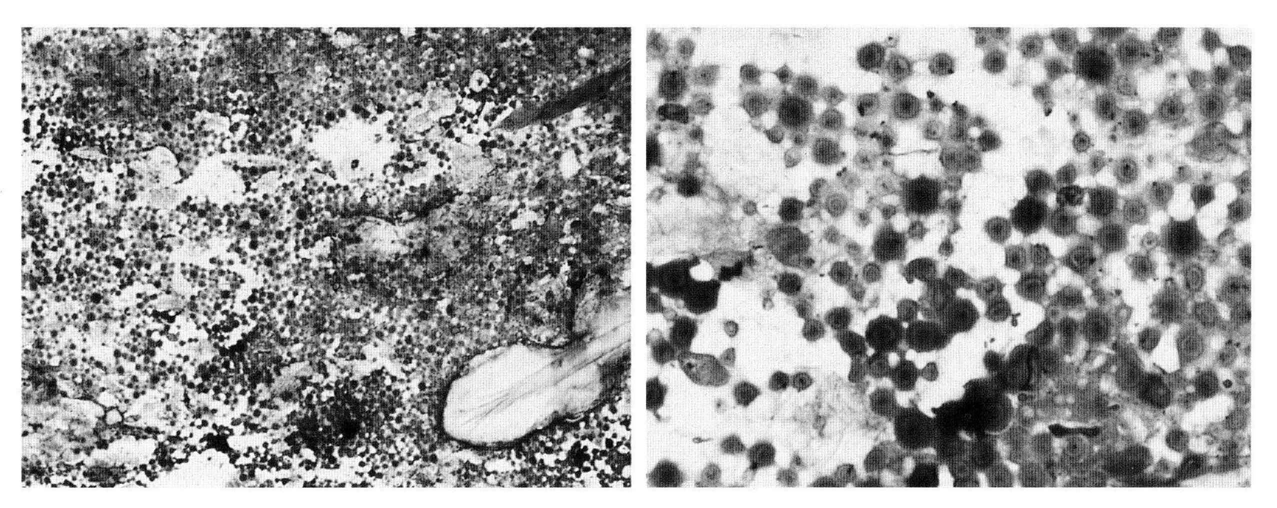

Figure 19.12 Unlike the rectangular viral inclusion bodies of molluscum contagiosum, the fungal organisms of *C. neoformans* are round, with a clear halo surrounding a darker central area (PMS fungal / Tzanck; left: ×50; right: ×150)

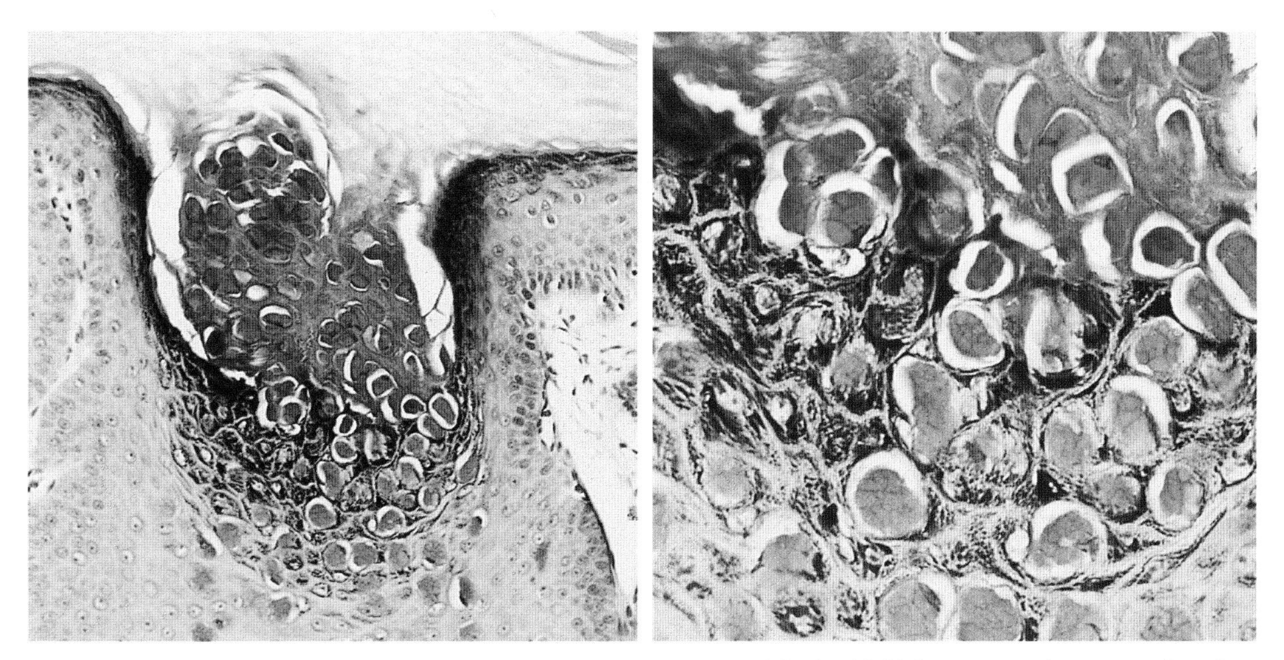

Figure 19.13 Skin biopsy of molluscum contagiosum shows eosinophilic (reddish) large, amorphous, round molluscum bodies in the basal layers of the epidermis whereas those in the upper layers of the epithelium are basophilic (bluish) in color (left). The molluscum bodies compress the nuclei of infected keratinocytes into crescent-shaped remnants, seen at the periphery of the cells (right). (H&E; left: ×50; right: ×100)

Biopsy

A tissue specimen for microscopic examination may be obtained either by curettage (using a curette), shave excision (using a scalpel with a number 15 blade) or punch biopsy. The pathological changes are distinctive and diagnostic. With H&E staining, the large round molluscum bodies in the basal layers of the epidermis are eosinophilic. However, in the upper layers of the epithelium, the amorphous viral inclusions appear basophilic (Figure 19.13). There is hyperplasia of the involved epidermis, and the nuclei of the infected keratinocytes are compressed into crescent-shaped remnants at the

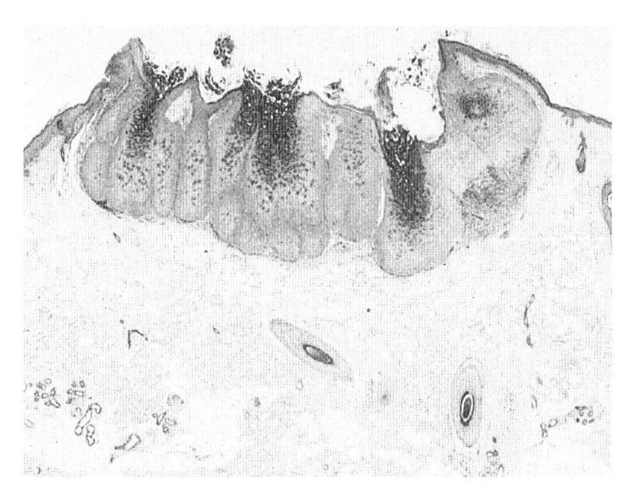

Figure 19.14 Skin biopsy of molluscum contagiosum shows hyperplasia of the epidermis. Numerous molluscum bodies are seen within the keratinocytes (H & E; ×5)

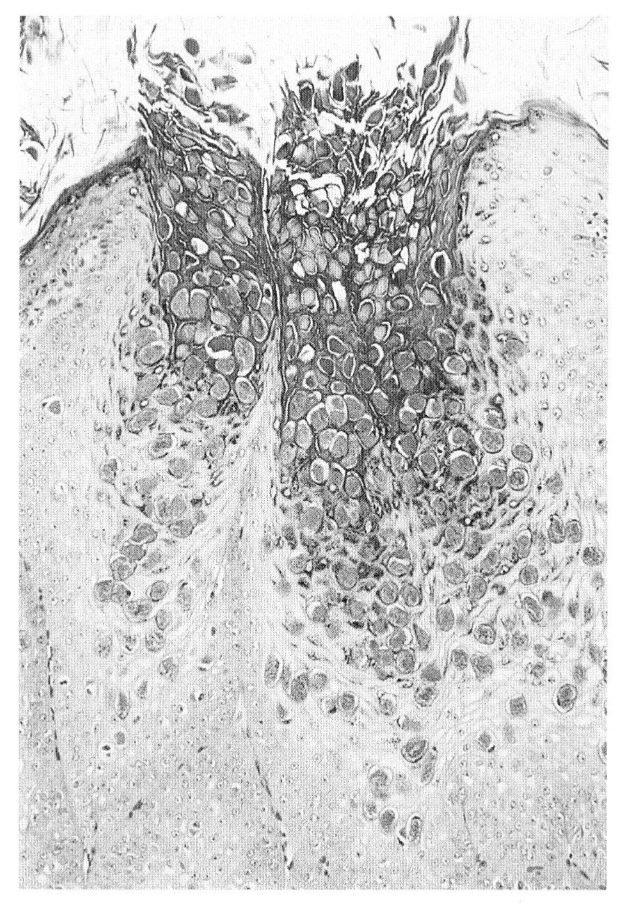

Figure 19.15 Higher magnification of Figure 19.14 shows discharge of viral inclusion bodies to the skin surface through central channels (H & E; ×25)

cell periphery by the molluscum bodies. Often, a central channel, through which the molluscum bodies are discharged to the surface, may be noted; clinically, this channel corresponds to the central core of the umbilicated papule (Figures 19.14 and 19.15).

Bibliography

Epstein WL. Molluscum contagiosum. *Semin Dermatol* 1992;11:184–9

Hicks MJ, Flaitz CM, Cohen PR. Perioral and cutaneous umbilicated nodular lesions in acquired immunodeficiency syndrome. *Oral Surg Oral Med Oral Pathol Oral Radiol Endod* 1997;83:189–91

Margolis S. Genital warts and molluscum contagiosum. *Urol Clin North Am* 1984;11:163–70

Shelley WB, Burmeister V. Office diagnosis of molluscum contagiosum by light microscopic demonstration of virions. *Cutis* 1985;December:465–6

Index

A

acne
- acne rosacea, *Demodex* mite implication 48
- acne vulgaris 28
- free fatty acids 28
- pathogenesis 27, 28

Acremonium 87–88
- clinical features 87
- source 87

acrochordons, molluscum contagiosum differentiation 120
alopecia, demodectic 48
Alternaria 92–93
- clinical features 92
- source 92

Amblyomma spp 50–51
- *A. americanum*
  - Lone Star tick 52
  - tularemia 52
- removal 51

anaerobiosis 12
- GasPak method 12–13

anthrax
- *Bacillus anthracis* 36
- cutaneous 36

antibiotic-sensitivity tests 13–14
- Bauer–Kirby method 13
- disk method 13
- mycobacteria 44
- procedure 13–14

antibiotics, *Pseudomonas aeruginosa* susceptibility 29
antideoxyribonuclease (DNase) B titer, streptococcal serological tests 22
antihyaluronidase titer, streptococcal serological tests 22
antimetabolites, *Pseudomonas aeruginosa* susceptibility 29
antinicotinamide-adenine dinucleotide (NAD)ase titer, streptococcal serological tests 22
antistreptolysin O (ASO) titer, streptococcal serological tests 22
'apple-jelly' nodules, lupus vulgaris 42
arachnids 48–55
arthroconidia 105
arthropods 45–66
- *Francisella (Pasteurella) tularensis* 37
- and human disease 46–47
- identification
  - clinical indications 46
  - equipment 46
  - interpretation 47
  - methods and general principles 46–47
  - procedure 46–47
  - procedure, operating 46–47
  - test principles 46

Ascaris lumbricoides 66
aspergillosis 86
Aspergillus spp 86–87
- *A. flavus* 86
- *A. fumigatus* 86, 87
- *A. niger* 86
- *A. terreus* 86
- clinical features 86–87
- source 86

automated reagin test (ART), syphilis 37
Automeris io, io moth caterpillar 62–63

B

Bacillus anthracis 36
- culture 36
- direct smear 36

morphology 36
specimen collection 36
bacitracin disk test, streptococcal differentiation 21
bacteria 9–44
bacterial culture 11–12
bacterial culture media 11–13
bacterial metabolites 17
bartonellosis, Psychodidae sandfly 60
basal cell carcinoma, molluscum contagiosum differentiation 120
Bauer–Kirby method, antibiotic-sensitivity tests 13
bedbugs 60
bees 61
allergy, epinephrine 61
beetles, blister 59
biopsy
CMV 118–119
HSV and VZV infections 118
Bipolaris
clinical features 93
source 93
biting cattle lice 58–59
black fungi 86, 92–97
black widow spider, bite treatments 53
blastoconidia 68, 69
Blastomyces dermatitidis 103–104
clinical features 103
source 103
blastomycosis 103
body lice 58
borrelial relapsing fever, *Ornithodoros* 52–53
boutonneuse fever, *Rhipicephalus sanguineus* 52
Bovicola bovis, biting cattle lice 58–59
brown dog tick 52
brown recluse spiders 53
brucellae, culture incubation 12
bugs 60
burns, CMV infection 118
butterflies 62–63

C

C carbohydrates, β-hemolytic streptococci subdivision 20
cancer, CMV infection 118
Candida spp 98–99
C. albicans 98–99
clinical features 98
paronychia 30
source 98
C. glabrata 98
C. guilliermondi 98
C. krusei 98
C. lusitaniae 98
C. parapsilosis 98
C. pseudotropicalis 98
C. tropicalis 98
carbuncles 18
Caripito itch, *Hylesia* moths 62
cat-scratch disease, cat flea transmission 63, 64
catalase test
staphylococci 17
streptococci 21
cellulitis 23
streptococcal culture 21
centipedes 56
Cephalosporium see Acremonium
cervical lymphadenitis, *M. scrofulaceum* 43
Cheyletiella mites, pet-associated dermatosis 49
chiggers 49
chigoe flea 63
Chilopoda and Diplopoda 56
chocolate blood agar 13
Chorioptes, astigmata mites 50
chromoblastomycosis
Cladosporium carrionii 94
Fonsecaea 95
Phialophora verrucosa 96
Rhinocladiella aquaspersa 97
sclerotic bodies 68, 69
Cimex spp
C. hemipterus 60
C. lectularius 60
Cladosporium spp 94
C. carrionii 94
clinical features 94
source 94
Cladosporium werneckii 95–96
Clostridium 36
CMV see cytomegalovirus
coagulase tests
catalase test 17
phage typing 17
slide coagulase test 17
staphylococci, pathogenicity 17
tube coagulase test 17
Coccidioides immitis 104–105
clinical features 104–105
source 104
coccidioidomycosis
Coccidioides immitis 104
spherules 68, 69
Coleoptera, blister beetles 59
Colorado tick fever 51
Columbicola columbae, pigeon lice 58–59

Cordylobia anthropophaga, tumbu flies 59
corticosteroids, *Pseudomonas aeruginosa* susceptibility 29
corynebacteria 25–27
  Albert's stain 25
  Corynebacterium diphtheriae 22, 25
  C. minutissimum, culture 26
  C. tenuis 25
  diphtheroid rods 25
  erythrasma 25–26
  pitted keratolysis 27
  trichomycosis 26–27
crab lice 57
Cryptococcus neoformans 99–100
  clinical features 99–100
  molluscum contagiosum differentiation 122–123
  source 99
Ctenocephalides felis, cat flea 63, 64
Culicidae, viral and helminthic pathogen transmission 60
culture media *see* media
Curvularia 94–95
  source 94
Cuterebra flies, human myiasis 59
cytomegalovirus 118–119
  biopsy 118–119
    'owl-eyes inclusions' 119
  clinical presentation 118
  viral culture 118

D

dematiaceous molds 68, 86, 92–97
Demodex spp, mites
  D. brevis 48
  D. folliculorum 48, 49
Dermacentor spp, ticks 50–51
  D. andersoni
    ehrlichiosis 51
    tularemia 51
    wood tick 51
  D. variabilis
    dog tick 52
    tularemia 52
  removal 51
Dermatobia hominis 59
Dermatophagoides mites 50
Dermatophilus congolensis 27
dermatophytes
  clinical disease 72
  epidemiology 71–72
  morphology *in vivo / in vitro* 72–73
  nutritional requirements 73
  pathogenic species 71
  sporulation characteristics 73
dermatophytosis
  KOH-positive 32
  specimen collection 72
  see also tinea
Dipetalonema (Acanthocheilonema, Mansonella) worms, midge transmission 60
diplococci, gram-negative intracellular, gonorrhea 34
Diplopoda and Chilopoda 56
Diptera 59–60
diptheroid, defined 25
dog tick 52

E

EBV *see* Epstein–Barr virus
Echidnophaga gallinacea, stick-tight flea 63
Echinolaelaps (Laelaps), rodent mites 50
ecthyma gangrenosum, malignant external 30
ectoparasites, arthropods and pinworms 45–66
ehrlichiosis
  Dermacentor andersoni 51
  Rhipicephalus sanguineus 52
endocarditis, *Rhodotorula* 101
Enterobacter spp 30, 30–33
  culture 30–31
  E. aerogenes, eosin / methylene blue (EMB) agar 31
  infection 31–32
  specimen collection 31–32
Enterobius vermicularis 65
eosin / methylene blue (EMB) agar
  Enterobacter aerogenes 31
  Escherichia coli culture 31
  Klebsiella spp 31
  Proteus culture 31
Epidermophyton floccosum 79–80
  clinical features 79
  source 79
epinephrine, bee sting allergy 61
Epstein–Barr virus 110
erysipelas 24
  streptococcal culture 21
erythema multiforme, tularemia differentiation 38
erythrasma
  Corynebacterium minutissimum 25
  culture 26
  direct examination 25
  infection 26
  toe-web site 26
erythrogenic toxin, scarlet fever rash 22
Escherichia coli 30

eosin/methylene blue (EMB) agar 31
exanthem, rubelliform type 118
Exophiala werneckii 95–96

F

fibroblast cells, viral culture 117
fire ants 61
fleas 63–64
flies 59–60
black 60
larvae 59
tsetse 60
flocculation test, syphilis 37
fluorescent treponemal antibody absorption (FTA-ABS) test, syphilis 37
folliculitis 17–18
culture procedures 32
Enterobacter spp 31–32
Klebsiella spp 31–32
Malassezia furfur 100, 101
Proteus spp 32
Fonsecaea 95
clinical features 95
source 95
foot, sole, infection 32
Francisella (Pasteurella) tularensis 37–38
culture 38
direct examination 38
infection 38
specimen collection 37
free fatty acids, acne 28
fungemia
Malassezia furfur 100
Rhodotorula 101
fungi 67–107
culture 70
dematiaceous molds 68, 86, 92–97
dermatophytes 71–85
see *also* dermatophytes
dimorphic pathogens 103–107
specimen collection 103
thermodimorphic 103
tissue dimorphic 103
direct microscopy 68–69
clearing solutions 69
staining agents 69
histopathology 69–70
hyphae 68, 69
monomorphic 68
thermodimorphic 68
see *also* molds
furuncles 18
Fusarium 88
clinical features 88
source 88

G

GasPak system
anaerobiosis 12–13
Propionibacterium acnes 28
Geotrichum 89
clinical features 89
source 89
glomerulonephritis
acute, streptococcal skin infections 22
acute hemorrhagic, scarlet fever complication 23
Glycyphagus mites 50
gonococci
culture incubation 12
human parasites 34
gonorrhea, gram-negative intracellular diplococci 34
Gram-negative bacteria 29–33
biochemical reactions 32
Gram's staining, methods 10–11
grape-like odor, *Pseudomonas aeruginosa* 29
green nail syndrome, *Pseudomonas aeruginosa* 30
grocer's itch mite, *Glycyphagus* 50
gypsy moth caterpillars 62

H

hair perforation test, dermatophyte identification 73
head lice 57–58
helminthic pathogens 65–66
Hemiptera (true bugs) 60
Henderson–Patterson bodies, molluscum contagiosum 120
herpes simplex and varicella–zoster virus 110–118
biopsy 118
clinical presentation 110
differentiation 110
Tzanck smear preparation 110–114
viral culture 115–117
herpesviruses 110–119
pathognomonic findings 114–115
Histoplasma capsulatum 105–106
clinical features 105–106
source 105
histoplasmosis, *Histoplasma capsulatum* 105
hot-tub folliculitis, *Pseudomonas aeruginosa* 30
house dust mite 50
HSV see herpes simplex virus
human immunodeficiency virus (HIV)
CMV infection 118
cryptococcosis lesions, molluscum contagiosum

differentiation 120, 122
hyaline molds 86–92
Hylesia moths, Caripito itch 62
Hymenoptera 61
hyphae
cenocytic 68, 69
pseudohyphae 69
septations 68

I
iatrogenic immunosuppression, CMV infection 118
immunosuppressive agents, *Pseudomonas aeruginosa* susceptibility 29
impetigo 18
streptococcal infection 21, 23
incubation 12
laboratory methods 12
Insecta 57–64
io moth caterpillar 62–63
Italian asp 62–63
Ixodes ticks, Lyme disease 51

J
jigger 63

K
keratolytic enzymes, trichomycosis 26
keratomycosis, *Scopulariopsis* 92
kissing bugs 60
Klebsiella spp 30
eosin / methylene blue (EMB) agar 31
K. pneumoniae, sheep's blood agar 31

L
Laelaps spp 50
L. castroi, mesostigmata mite 50
L. echidnina, rat mite 50
Latrodectus mactans, black widow spider 53
Leishmania mexicana, *Lutzomyia* transmission 60
leishmaniasis, fly transmission 60
Lepidoptera 62–63
leukonychia mycotica 86
Trichophyton mentagrophytes 75
lice
Anoplura and Mallophaga 57–59
body lice 58
lichen planus, molluscum contagiosum differentiation 120
lichen scrofulosus, tuberculosis 42
Liponyssoides (formerly *Allodermanyssus*), mouse mite 50
Lone Star tick 52
louse-borne relapsing fever, body lice 58
Löwenstein–Jensen medium
M. tuberculosis 40
mycobacteria 41
Loxosceles reclusa, brown recluse spider 53
Ludwig's angina, scarlet fever 23
lupus vulgaris
'apple-jelly' nodules 42
tuberculosis of the skin 42
Lutzomyia spp, *Leishmania mexicana* 60
Lymantria dispar, tent caterpillars 62
Lyme disease, *Ixodes* ticks 51

M
M protein, immunity 20
MacConkey agar, *M. fortuitum* culture 41–42
maculae ceruleae, pubic lice 57
maggots, human myiasis 59
Majocchi's granuloma 74
Malassezia furfur 100–101
clinical features 100–101
source 100
Mallophaga (suborder Amblycera) lice 59
mange
human, *Demodex* mites 48
human infestation 50
mannitol fermentation, *Staphylococcus aureus* 16
media
bacteria culture 11–13
Löwenstein–Jensen medium, mycobacteria 41
oxidation–fermentation *P. aeruginosa* 29
primary, differential and selective 13
sample collection and culture, tuberculosis 40
viral culture, transport media 117
Megalopyge opercularis 62–63
meningitis
Acremonium 87
Rhodotorula 101
meningococci, culture incubation 12
mesostigmata mites 50
Micrococcus sedentarius 27
Microsporum spp
M. canis 80–81
clinical diagnosis 72
clinical features 80
source 80
M. gypseum 71, 81–82
clinical features 81
source 81
millipedes 56
mites
astigmata 50

Cheyletiella 49
Demodex folliculorum 48, 49
grocer's itch mite 50
house dust 50
mesostigmata 50
Laelaps castroi 50
prostigmata 49–50
Sarcoptes scabiei 48
Trombidiidae 49
zoonotic 48–49
molds
dematiaceous 68, 86, 92–97
morphology 68, 69
non-dermatophytic 86–92
molluscum contagiosum 120–124
biopsy 123–124
clinical presentation 120
diagnosis 120–123
monoclonal antibodies, *Bacillus anthracis* identification 36
mosquitoes, viral and helminthic pathogen transmission 60
moths 62–63
mouse mite, *Liponyssoides* (formerly *Allodermanyssus*) 50
Mucor 89–90
clinical features 89–90
source 89
mycetoma
Acremonium 87
Curvularia 94
Fusarium 88
Scedosporium apiospermum 91
mycobacteria 39–44
antibiotic-sensitivity testing 44
atypical bacteria, Runyon's classification 41–42
dysgonic 41
eugonic 41
M. africanum 40
M. avium–intracellulare complex (MAI or MAC)
culture 41
lesions 43
M. bovis 40
culture 41
M. chelonae culture 41–42
M. fortuitum culture 41–42
M. fortuitum–M. chelonae complex, lesions 43
M. haemophilum skin lesions 43
M. kansasii culture 41
M. leprae 39–40
direct smear 44
infection 44
lesions 44
M. marinum (*M. balnei*)
culture 41
lesions 42–43
M. scrofulaceum
culture 41
lesions 43
M. szulgai, culture 41
M. tuberculosis 40
culture 41
disseminated infection 42
lesions 42
localized infection 42
M. ulcerans
culture 41
lesions 43
Runyon's classification 39–40
mycology, laboratory methods 68–70
mycoses
Acremonium 87
Candida albicans 98
Cryptococcus neoformans 99
Fusarium 88
pulmonary, *Rhodotorula* 101
systemic
Rhodotorula 101
Trichosporon spp 102
myiasis
Cuterebra flies 59
fly larvae 59

N

Neisseria spp 34–35
culture 34–35
direct smear 34
infection 35–36
N. gonorrhoeae 34
culture, chocolate blood agar medium 13
identification 35
N. meningitidis 34
nits 57, 58
Norwegian (crusted) scabies 48

O

ocular mycoses
Acremonium 87
Fusarium 88
Onchocerca volvulus, black fly transmission 60
one-hand-two-foot syndrome 74
onychomycosis
Acremonium 87
Candida albicans 98

Fusarium 88
non-dermatophytic, *Scopulariopsis* 92
subungual 79
Trichosporon spp 102
opossum bug 62–63
Ornithodoros, borrelial relapsing fever 52–53
Ornithonyssus bacoti, rat mite 50
otitis, malignant external, *Pseudomonas aeruginosa* 30
Otodectes, astigmata mites 50
'owl-eyes inclusions', CMV biopsy 119
oxidation–fermentation medium, *Pseudomonas aeruginosa* 29

P

paracoccidioidomycosis 106
Paracoccidioides brasiliensis 106
clinical features 106
source 106
paronychia
Candida albicans 30
green nail syndrome 30
Pediculus humanus capitis, head lice 57–58
Pediculus humanus corporis, body lice 58
pemphigus neonatorum, malignant *see* staphylococcal scalded skin syndrome (SSSS)
Penicillium spp 90
clinical features 90
P. marneffei 90
source 90
peritonsillar abscess, scarlet fever 23
pet-associated dermatosis 49
Phaeoannellomyces werneckii 95–96
clinical features 95–96
source 95
phaeohyphomycosis
Alternaria 92, 93
Bipolaris 93
Curvularia 94
Phialophora richardsiae 96
phage typing, staphylococci 17
Phialophora spp 96–97
clinical features 96–97
P. richardsiae 96
P. verrucosa 96
source 96
photochromogens, atypical bacteria 41
Phthiraptera 57
pinworms, ectoparasites and arthropods 45–66
pitted keratolysis 27
direct smear 27
infection 27
pityriasis versicolor
Malassezia furfur 100
microscopy, 'spaghetti and meatballs' 100
Pityrosporum spp
P. orbiculare 100
P. ovale 100
plague, rat flea transmission 64
plantar vesiculobullous eruptions, gram-negative disease 32
polymerase chain reaction (PCR), *M. tuberculosis* detection 40
Propionibacterium spp 27–28
collection and culture 28
P. acnes 27
P. avidum 27
P. granulosum 27
prostigmata mites 49–50
Proteus spp 30
culture 30–31
P. mirabilis, skin invasion 30
P. vulgaris, skin invasion 30
Pseudallescheria boydii 91
Pseudomonas spp
culture, pyocyanin synthesis 29
P. aeruginosa 29
culture 29–30
infection 30
morphology 29
P. cepacia 30
septicemia 30
Psocoptera, book lice 58–59
Psoroptes spp
mites 50
P. communis, livestock scab 50
Pthirus pubis 57
pubic (crab) lice 57
Pulex irritans, 'human flea' 63
pulmonary mycoses
Fusarium 88
Geotrichum 89
puss caterpillar 62–63
pyocyanin synthesis, *Pseudomonas* agar 29
pyoderma, molluscum contagiosum differentiation 120

Q

quinsy, scarlet fever 23

R

rapid plasma reagin circle card test (RPR-CT), syphilis 37
rat flea 64
rat mite
Laelaps echidnina 50

Ornithonyssus bacoti 50
reduviid bugs 60
rheumatic fever, acute, scarlet fever complication 23
Rhinocladiella aquaspersa 97
clinical features 97
source 97
Rhipicephalus sanguineus
boutonneuse fever 52
brown dog tick 52
Rhizopus 90–91
clinical features 90
source 90
Rhodotorula 101–102
clinical features 101–102
source 101
rice grain test, dermatophyte identification 73
Rickettsia spp
R. felis, cat flea transmission 64
R. tsutsugamushi 49
rickettsial pox, mouse mites 50
Ritter's disease *see* staphylococcal scalded skin syndrome (SSSS)
Rocky Mountain spotted fever 51, 52
rodents
Francisella (Pasteurella) tularensis 37
mite, *Echinolaelaps* (*Laelaps*) 50
Romaña's sign, American trypanosomiasis 60, 61

S

saddleback caterpillar 62–63
Sarcoptes scabiei 48
scab, livestock, *Psoroptes communis* 50
scabies
moth dermatitis differentiation 62
Norwegian (crusted) 48
Sarcoptes scabiei 48
scarlet fever 23
complications 23
Dick test 22
erythrogenic toxin rash 22
Schultz–Charlton reaction 22
Scedosporium apiospermum 91
clinical features 91
source 91
Schistosoma spp
ova 66
S. hematobium 66
S. japonicum 66
S. mansoni 66
Schultz–Charlton reaction, scarlet fever 22
Scopulariopsis 92
clinical features 92
source 92
scorpions 54
Centruroides exilicauda (*C. sculpturatus*) 54
scotochromogens, atypical bacteria 41
scrofuloderma, skin lesions 42–43
scrub typhus, *Trombicula alfreddugèsi* 49
seborrheic dermatitis, *Malassezia furfur* 100
septicemia, *Pseudomonas* 30
serological procedures
antideoxyribonuclease (DNase) B titer, streptococci 22
antihyaluronidase titer, streptococci 22
antinicotinamide-adenine dinucleotide (NAD)ase titer, streptococci 22
antistreptolysin O (ASO) titer, streptococci 22
erythrogenic toxin, streptococci 22
syphilis 37
Sibine stimulea, saddleback caterpillar 62–63
Siphonaptera, fleas 63–64
sleeping sickness, African, *Trypanosoma gambiense* transmission 60
slide coagulase test, *Staphylococcus aureus* 17
smears, staining and cultivation
laboratory methods 10–14
smear preparation 10–11
Solenopsis spp
S. invicta 61
S. richteri 61
'spaghetti and meatballs', pityriasis versicolor microscopy 100
Spermophilus mexicanus 59
spiders 53–54
Latrodectus mactans, black widow spider 53
Loxosceles reclusa, brown recluse spiders 53
tarantula 53–54
Sporothrix schenckii 107
clinical features 107
source 107
sporotrichosis, *Sporothrix schenckii* 107
sporulation
Cladosporium type 96
Phialophora type 96
staphylococcal scalded skin syndrome (SSSS) 18
Tzanck smear 18
Staphylococcus spp 15–19
classification 15
coagulase-negative, infection 19
S. aureus 15–18
bacterial metabolites 17
catalase test 17
culture and identification 15–16
impetigo 18, 23

infection 17–18
morphology 15
specimen collection 15–16
S. epidermidis 18–19
colony morphology 18–19
infection 19
S. saprophyticus 19
staphylothrombin formation 17
starch agars, dermatophyte identification 73
stick-tight flea 63
streaking, technique 13–14
streptococci 20–24
classification 20
culture and identification 21–24
serological tests 22
infection 22–24
M protein immunity 20
morphology 20
specimen collection 20–21
Streptococcus pyogenes 20
swimming-pool granuloma, *M. marinum* 42–43
syphilis 36–37
fluorescent treponemal antibody absorption (FTA-ABS) test 37
see also Treponema pallidum

T

tarantulas 53–54
Theraphosidae, tarantulas 53–54
thrush-like disease, *Geotrichum* 89
ticks 50–53
Amblyomma spp 50
A. americanum, Lone Star tick 52
Boophilus 52
Dermacentor spp 50
D. andersoni, wood tick 51
D. variabilis, dog tick 52
hard (Ixodidae) 50
Ixodes scapularis, black-legged tick 51
Ornithodoros 52–53
removal 51
Rhipicephalus sanguineus, brown dog tick 52
soft (Argasidae) 50
tick paralysis 51, 52
tinea
classification 72
dermatophyte ecology 71
tinea capitis
black-dot
Trichophyton tonsurans 77
Trichophyton violaceum 83
ectothrix fluorescent, *Microsporum canis* 80
endothrix 84
Trichophyton tonsurans 78
favus, *Trichophyton schoenleinii* 83
tinea corporis 81
tinea nigra, *Phaeoannellomyces werneckii* 95–96
tinea pedis, vesicular, *Trichophyton mentagrophytes* 75, 76
tinea unguium 79
toeweb infection, *Pseudomonas aeruginosa* 30
trench fever, body lice 58
Treponema pallidum
direct examination 37
serological procedures 37
specimen collection 36–37
syphilis 36–37
Triatoma bugs, trypanosomiasis 60, 61
trichomycosis 26–27
culture 26
direct examination 26
infection 26–27
Trichophyton agars, dermatophyte identification 73
Trichophyton spp
T. equinum 84–85
clinical features 84
source 84
T. mentagrophytes 75–76
clinical features 75–76
source 75
T. rubrum differentiation 75
T. rubrum 74–75
clinical features 74–75
source 74
T. schoenleinii 83
clinical features 83
source 83
T. tonsurans 77–78
clinical diagnosis 72
clinical features 77
source 77
T. verrucosum 82
clinical features 82
source 82
T. violaceum 83–84
clinical features 83–84
source 83
Trichosporon spp 102
clinical features 102
source 102
T. beigelii 102
T. cutaneum 102
Trichuris trichiura 66
Trombicula alfreddugèsi 49

larval identification 49
Trypanosoma gambiense
African sleeping sickness 60
tsetse fly 60
trypanosomiasis, *Triatoma* bugs 60, 61
tsetse fly, *Trypanosoma gambiense* transmission 60
tuberculosis
morphology 40
sample collection and culture 40
media 40
skin lesions
lupus vulgaris 42
tuberculosis verrucosa 42
see also mycobacteria, *M. tuberculosis*
tularemia 37–38
Amblyomma americanum 52
Chrysops 60
Dermacentor andersoni 51
Dermacentor variabilis 52
see also Francisella (Pasteurella) tularensis
tumbu flies 59
Tunga penetrans, chigoe flea 63
typhus
endemic, cat flea 63–64
epidemic, body lice 58
Tzanck smear
HSV and VZV infections 110–114
SSSS 18

U

urease test, dermatophyte identification 73

V

vagabond's disease, body lice 58
varicella, molluscum contagiosum differentiation 120
varicella–zoster virus
HSV differentiation 110
see also herpes simplex virus
Venereal Disease Research Laboratory (VDRL) flocculation test 37
verruca
molluscum contagiosum differentiation 120
'stadium' edge plaques, *Blastomyces dermatitidis* 103
viral culture
CMV 118
diagnosing reagents in office laboratory 112
false-negative results 116–117
fibroblast cells 117
HSV infection 115–117
transport media 117
viruses 109–124
VZV *see* varicella–zoster virus

W

wasps 61
wheel bugs 60–61
white piedra, *Trichosporon* spp 102
white spot 75
wooly asp 62–63
wooly slug 62–63
worms 65–66

X

Xenopsylla cheopis, rat flea 64

Y

yeasts 98–102
morphology 98

Z

zoonoses
lice 58–59
mites 48–49
pet-associated dermatosis 49
zygomycosis
Mucor 89
Rhizopus 90, 91